THE
MOST DIFFICULT
LOGIC
PROBLEMS

Published in 2020 by Welbeck Non-Fiction
An imprint of Welbeck Publishing Group
20 Mortimer Street
London W1T 3JW

Text and Design © 2020 Welbeck Publishing Group

Editorial: Georgia Goodall
Design Manager: Stephen Cary

A CIP catalogue for this book is available from the British Library.

ISBN: 978-1-78739-631-9

All images: © iStockphoto & Shutterstock

Printed in Dubai

10 9 8 7 6 5 4 3 2 1

THE
MOST DIFFICULT
LOGIC
PROBLEMS

Test your powers of reasoning with hundreds of exacting enigmas

Dr Gareth Moore

WELBECK

CONTENTS

INTRODUCTION

Welcome to *The Most Difficult Logic Problems*, packed from cover to cover with 200 extra-challenging puzzles of 20 highly varied types. Despite all of the puzzles taking the difficulty level up to 10, they are still designed to be solved – smart thinking and occasional inspiration should be all you need to crack every one of the problems.

The chances are that you won't have come across several of the puzzle types in this book, and even for those that seem relatively familiar you'll find yourself tackling them in new and extra-fiendish forms, such as in the warped logic of 'Wraparound Sudoku', or the twisted challenge of 'Double Minesweeper'. These will add to the difficulty, since you'll need to gain familiarity with the particular logic of each type of puzzle. The good news, however, is that your brain loves learning new skills, so as long as you're enjoying it then you'll also be racking up the mental benefits.

Some of the puzzles might take a bit of practice to get the hang of, but try not to skip over the ones that you find extra tricky – these are probably the ones that will benefit you the most. Your brain thrives on new challenges, and the more novel and unusual a task is, the better.

If there are any puzzles where the rules aren't clear to you, go ahead and check one of the answers at the back to see what a solved puzzle looks like. You can use this to work out how the given clues connect with the resulting solution, which can be helpful for understanding some of the more complex conundrums.

When you get stuck – and you will – remember that it's okay to make an intelligent guess, or even just a wild stab in the dark! Children learn by experimenting, and it turns out this works really well for these types of logic puzzles too. It might even be the quickest way to learn how to solve some of them. Your brain is great at spotting patterns, so jump in, try something and see what happens – even if it doesn't work out, you'll still discover things about how the puzzle works, helping you to make your next guess even smarter.

The puzzles are all presented from a 3D perspective, adding some additional visual reasoning elements to the puzzles. One effect of this dramatic presentation is that any references to "horizontal" or "vertical" in the instructions should be taken with regard to the angle the puzzle is drawn at, so if it is slanting down to the right then any "horizontal" lines will slant down to the right too. This might sound confusing, but all will become clear when you look at the relevant puzzles. A similar effect applies to any references to "rows" and "columns" too, where they will also follow the 3D perspective of the puzzle.

Finally, while your brain does indeed love novelty, it's important to also note that it doesn't learn well when frustrated. If you get truly stuck on a puzzle then don't be afraid to take a look at the solutions to 'borrow' a few extra clues to get you going again. It's better to complete the puzzle with help than to abandon it entirely.

Most of all, remember to have fun!

– **Dr Gareth Moore**, London

PUZZLES

BLACKOUT SQUARE

01 Place a number from 1 to 8 into each square, so no number repeats within any row or column of the grid. Beware: each empty square may represent two different 'missing' numbers – one for the row, and one for the column.

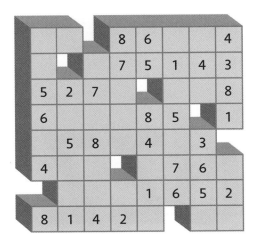

SOLUTION page 112

02

SOLUTION page 114

DOUBLE MINESWEEPER

03 Place one or two mines in some of the empty squares so that every given number has that many mines in its touching squares, including diagonally.

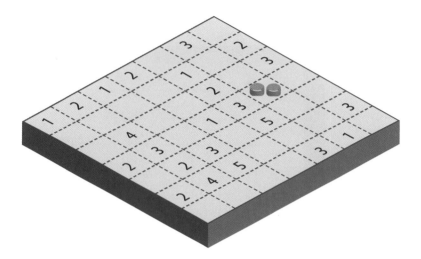

SOLUTION page 112

 04

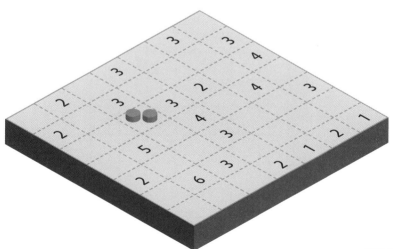

SOLUTION page 114

MYSTERY CALCUDOKU

05 Place a number from 1 to 8 once each into every row and column of the grid, while obeying the region clues. The value at the top left of each bold-lined region must be obtained when all of the numbers in that region have one of the four operations +, -, × and ÷ applied between them. To calculate - and ÷ results, begin with the largest number in the region and then subtract or divide by the other numbers in the region in any order.

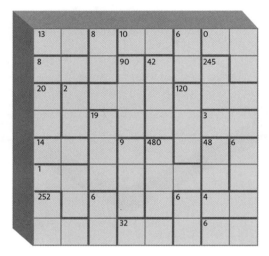

SOLUTION page 112

06

SOLUTION page 114

LINESWEEPER

07 Draw a loop which travels horizontally and vertically through the centres of some of the empty squares. The loop must pass by each numbered square in the grid the given number of times. These counts represent the number of side touching and diagonally touching squares visited by the loop.

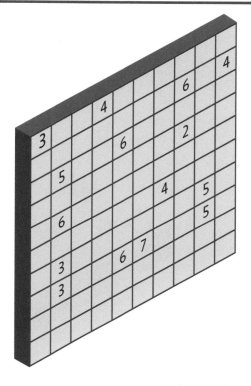

SOLUTION page 112

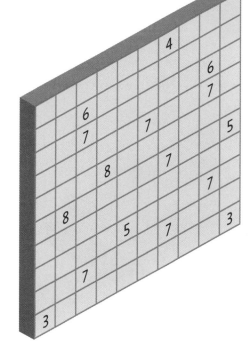

SOLUTION page 114

SHAPELINK DIAGONAL

 Draw paths to join pairs of matching shapes. Paths can travel horizontally, vertically or at a 45-degree angle, and no more than one path can enter any square. Paths cannot cross, except diagonally on the intersection of two gridlines.

SOLUTION page 112

SOLUTION page 114

HASHI

 Draw horizontal and vertical lines to join circled numbers. Each circle contains a number which specifies the number of lines that connect to it. No more than two lines may join any pair of circles. Lines may not cross other lines or circles. All circles must be joined in such a way that you can travel from any circle to any other circle by following one or more lines.

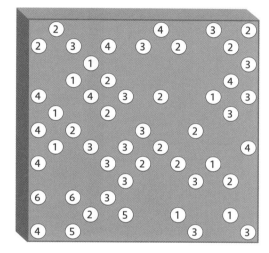

SOLUTION page 112

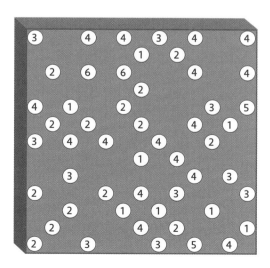

SOLUTION page 114

SPIRAL GALAXIES

13 Draw along some of the grid lines in order to divide the grid up into a set of regions. Every region must contain exactly one circle at its centre, and the region must be symmetrical in such a way that if rotated 180 degrees around the circle it would look exactly the same.

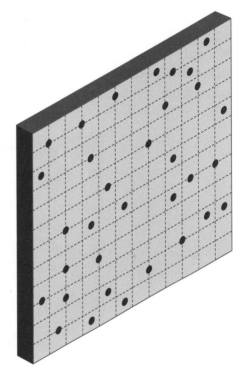

SOLUTION page 113

14

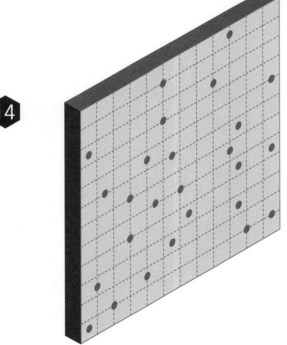

SOLUTION page 115

TOILS

15 Shade squares on the grid in order to place all five of the tetrominoes (T, O, I, L and S as shown) once each into the grid, so that every numbered row or column contains that many shaded squares. Tetrominoes can be rotated or reflected, but may not touch one another – except diagonally.

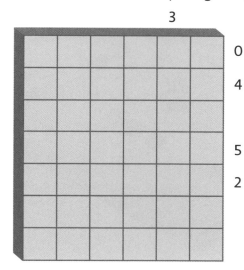

3

0
4

5
2

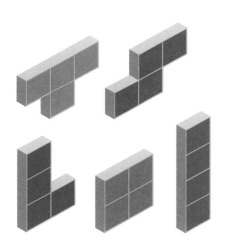

SOLUTION page 113

16

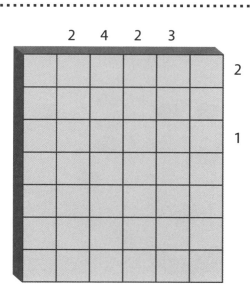

2 4 2 3

2

1

SOLUTION page 115

TENTS

17 Attach one tent to each tree, by placing it in a touching square in the same row or column. Numbers outside the grid reveal the total number of tents in certain rows and columns. Tents cannot touch each other – not even diagonally.

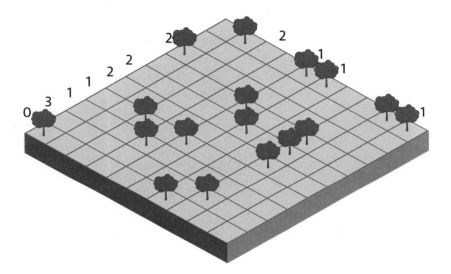

SOLUTION page 113

18

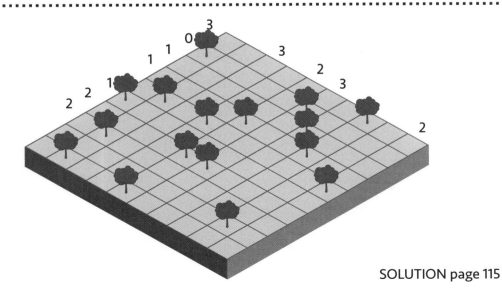

SOLUTION page 115

KING'S JOURNEY

 19 Place a number into every empty square so that the numbers 1 to 100 appear once each in the grid. Also, a path from 1 to 100 must be formed which could be travelled step-by-step as a king moves in chess – i.e. starting at 1, the path must travel to 2, 3, 4, 5 and so on while moving only to side touching or diagonally touching squares.

37	36		21	18	15	14			10
	39					13	11		
	41	34	31					7	1
	46	33		30	28	26	100		2
	45	47					97	5	
		51	50			98	96		
			58	61		93		86	
54	55			92	91	88	87		
69			73			79	83		
	67		72						82

SOLUTION page 113

20

					45	22	20		17
62	60	64			44	23			
69	67	65			48				13
70				49		25		14	
73		54				9	6		
			51						33
79									
							100		
	82	81	1		90	92	99	95	97
83				87	89				

SOLUTION page 115

NUMBER DARTS

21 Can you make each of the totals shown? For each total, choose one number from each of the four rings, so that those four numbers add to the given total.

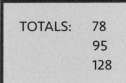

TOTALS: 78
95
128

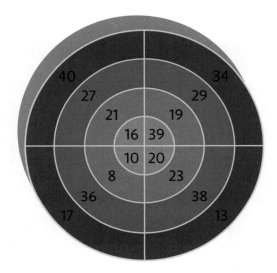

SOLUTION page 113

22

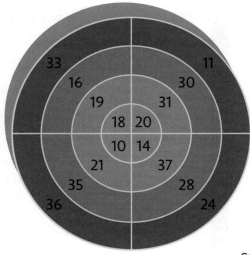

TOTALS: 64
71
125

SOLUTION page 115

LIGHTHOUSES

 23 Place a set of single-square ships in the grid. The number on each lighthouse reveals the total number of ships found in its row and column. Ships cannot touch either themselves or a lighthouse – not even diagonally. Ships must be in sight of a lighthouse, so may only be placed in a row or column which contains at least one lighthouse.

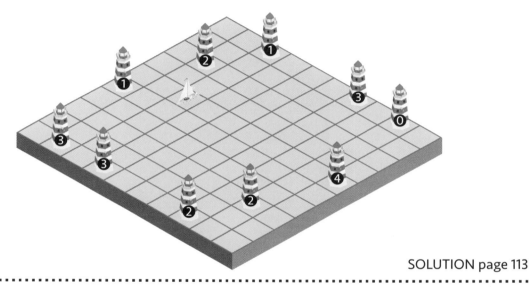

SOLUTION page 113

 24

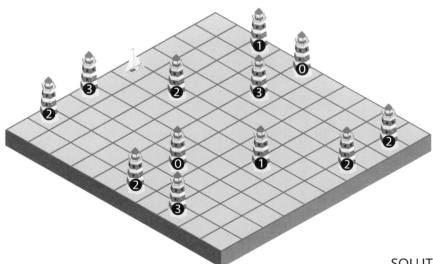

SOLUTION page 115

EASY AS A, B, C, D

25 Place a letter from A to D into some of the squares on the grid, so that each letter appears once in every row and column. This means that there will also be two empty squares in each row and column. Letters outside the grid reveal the first letter encountered in that row or column, from the viewpoint of the clue letter.

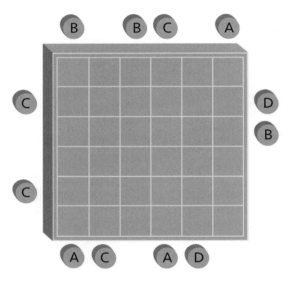

SOLUTION page 116

26

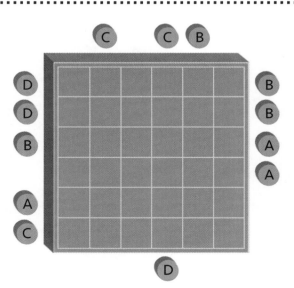

SOLUTION page 117

ARROW SUDOKU

27 Place numbers from 1 to 9 so that each number appears once in every row, column and 3x3 box. Also, each circled number must be equal to the sum of the numbers along its attached arrow.

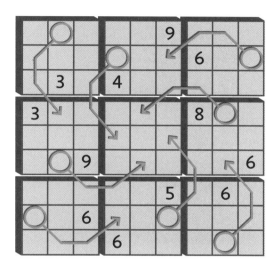

SOLUTION page 116

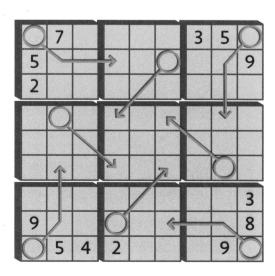

SOLUTION page 117

CLOUDS

 29 Shade some squares and/or rectangles of at least 2x2 in area, to form a set of clouds. Numbers outside the grid reveal the total number of shaded squares in each row and column. Clouds cannot touch – not even diagonally.

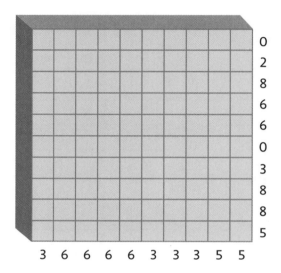

0
2
8
6
6
0
3
8
8
5

3 6 6 6 6 3 3 3 5 5

SOLUTION page 116

30

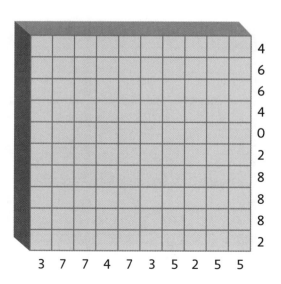

4
6
6
4
0
2
8
8
8
2

3 7 7 4 7 3 5 2 5 5

SOLUTION page 117

ARROWS

31 Place an arrow in each box outside the grid so that each number inside the grid has the given number of arrows pointing at it. Arrows can point up, down, left, right or diagonally. All arrows must point to at least one number.

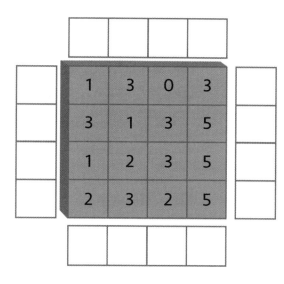

SOLUTION page 116

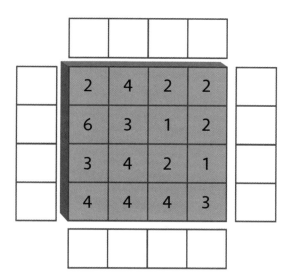

SOLUTION page 117

NO FOUR IN A ROW

33 Place either an 'X' or an 'O' into every empty square, so that no lines of four or more 'X's or 'O's are made in any direction in the grid, including diagonally.

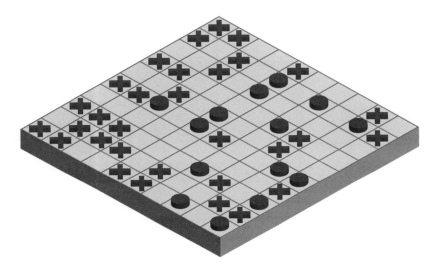

SOLUTION page 116

SOLUTION page 118

WRAPAROUND SUDOKU

35 Place a letter from 1 to 9 into each empty square so that every letter appears once in each row, column and bold-lined jigsaw shape. Some jigsaw shapes 'wrap around' from one side of the grid to the other, continuing at the opposite end of the same row or column.

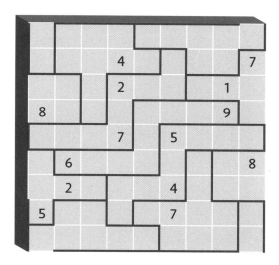

SOLUTION page 116

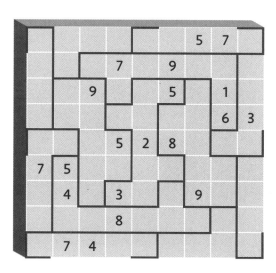

SOLUTION page 118

SKYSCRAPERS

 Place a digit from 1 to 7 into every square, so that no digit repeats in any row or column inside the grid. Place digits in such a way that each given clue number outside the grid represents the number of digits that are 'visible' from that point, looking along that clue's row or column. A digit is visible if there is no higher digit preceding it.

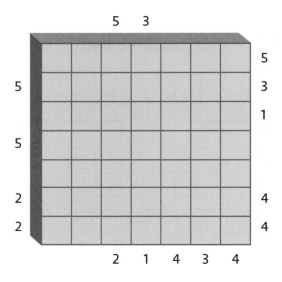

SOLUTION page 117

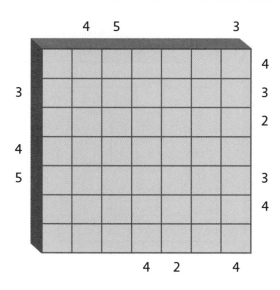

SOLUTION page 118

TRAIN TRACKS

 39 Complete the railway track so that it enters and exits the grid only where shown. When the track enters a square, it can either turn 90 degrees or pass straight through the square. It cannot cross over itself. Numbers outside the grid reveal the total number of squares containing track in each row and column. Rows or columns without a number may contain any number of track segments.

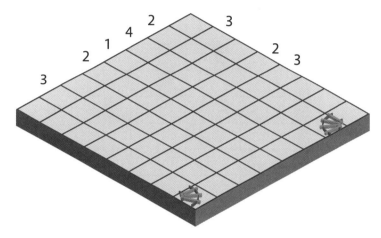

SOLUTION page 117

 40

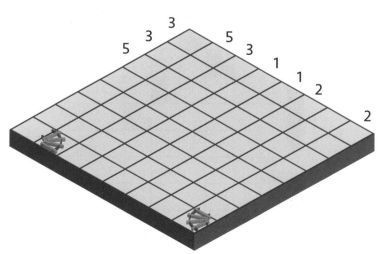

SOLUTION page 118

29

BLACKOUT SQUARE

41 Place a number from 1 to 8 into each square, so no number repeats within any row or column of the grid. Beware: each empty square may represent two different 'missing' numbers – one for the row, and one for the column.

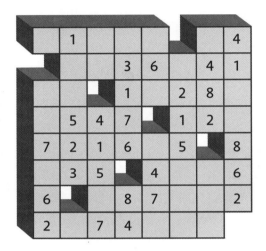

SOLUTION page 118

SOLUTION page 120

DOUBLE MINESWEEPER

43 Place one or two mines in some of the empty squares so that every given number has that many mines in its touching squares, including diagonally.

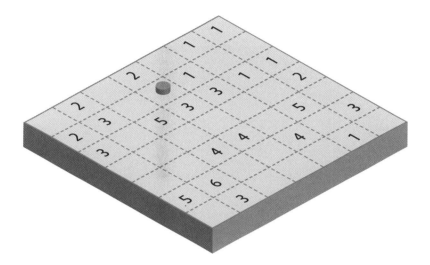

SOLUTION page 118

44

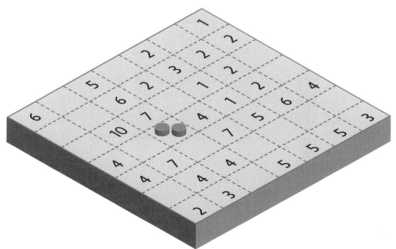

SOLUTION page 120

MYSTERY CALCUDOKU

 45 Place a number from 1 to 8 once each into every row and column of the grid, while obeying the region clues. The value at the top left of each bold-lined region must be obtained when all of the numbers in that region have one of the four operations +, -, × and ÷ applied between them. To calculate - and ÷ results, begin with the largest number in the region and then subtract or divide by the other numbers in the region in any order.

11		56		1		6	
	21		24	56		9	
0	48			12			13
	20				9		
	10		6	18	56		
	160	12			24		
					16		224
7		30		1			

46

21	32	2			7	2	
		15		6		14	9
			17		50		
35	6			168			12
	84				9		
32	7		7		19		
		28		24		21	
2			90				

SOLUTION page 121

LINESWEEPER

47 Draw a loop which travels horizontally and vertically through the centres of some of the empty squares. The loop must pass by each numbered square in the grid the given number of times. These counts represent the number of side touching and diagonally touching squares visited by the loop.

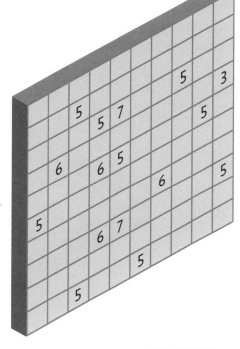

SOLUTION page 119

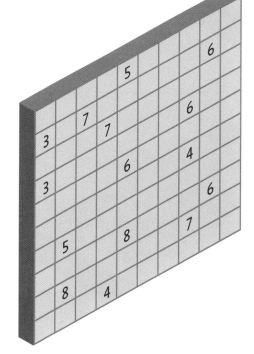

SOLUTION page 121

SHAPELINK DIAGONAL

49 Draw paths to join pairs of matching shapes. Paths can travel horizontally, vertically or at a 45-degree angle, and no more than one path can enter any square. Paths cannot cross, except diagonally on the intersection of two gridlines.

SOLUTION page 119

 50

SOLUTION page 121

HASHI

51 Draw horizontal and vertical lines to join circled numbers. Each circle contains a number which specifies the number of lines that connect to it. No more than two lines may join any pair of circles. Lines may not cross other lines or circles. All circles must be joined in such a way that you can travel from any circle to any other circle by following one or more lines.

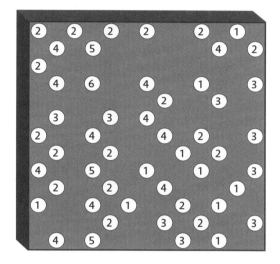

SOLUTION page 119

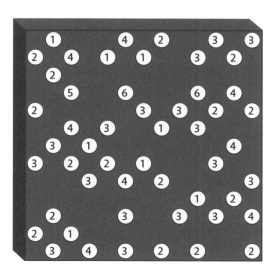

SOLUTION page 121

SPIRAL GALAXIES

53 Draw along some of the grid lines in order to divide the grid up into a set of regions. Every region must contain exactly one circle at its centre, and the region must be symmetrical in such a way that if rotated 180 degrees around the circle it would look exactly the same.

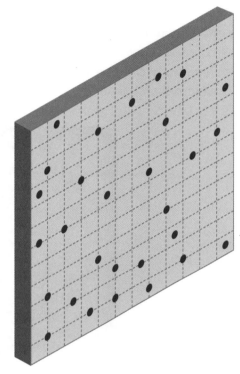

SOLUTION page 119

54

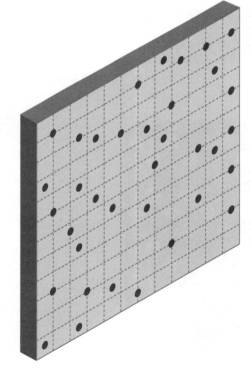

SOLUTION page 121

TOILS

55 Shade squares on the grid in order to place all five of the tetrominoes (T, O, I, L and S as shown) once each into the grid, so that every numbered row or column contains that many shaded squares. Tetrominoes can be rotated or reflected, but may not touch one another – except diagonally.

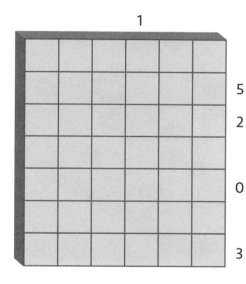

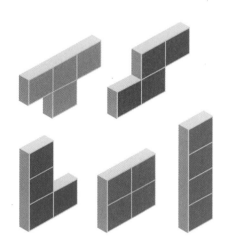

SOLUTION page 119

 56

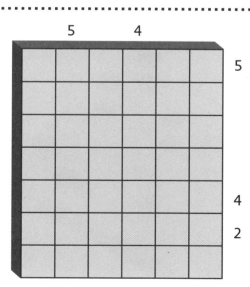

SOLUTION page 121

TENTS

 57 Attach one tent to each tree, by placing it in a touching square in the same row or column. Numbers outside the grid reveal the total number of tents in certain rows and columns. Tents cannot touch each other – not even diagonally.

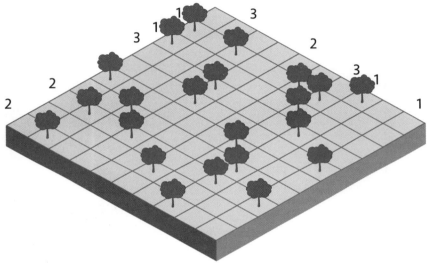

SOLUTION page 120

 58

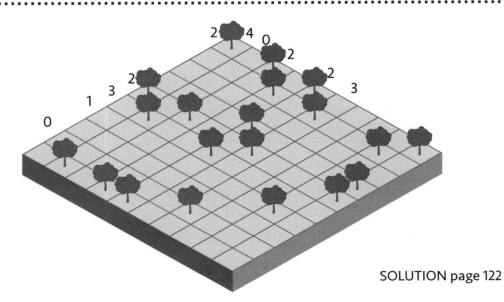

SOLUTION page 122

KING'S JOURNEY

 Place a number into every empty square so that the numbers 1 to 100 appear once each in the grid. Also, a path from 1 to 100 must be formed which could be travelled step-by-step as a king moves in chess – i.e. starting at 1, the path must travel to 2, 3, 4, 5 and so on while moving only to side touching or diagonally touching squares.

59

							68	69	1
78	76	80	81	82					2
		89		87		71		3	
	90	96	88						4
	92			100			24	23	
		59	60					10	
	54				28	30		11	
								13	
	44	50	42			34			14
47		45			38	37	36	17	15

SOLUTION page 120

60

35							24	18	17
38			32		29		20		16
		37	98	100			4		14
40	96		90		1				13
	94			88		6			11
					86	8	9		67
44	43				85	60	71	69	
49							61	70	
	50	82				74	77		
			81						

SOLUTION page 122

NUMBER DARTS

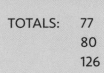

61 Can you make each of the totals shown? For each total, choose one number from each of the four rings, so that those four numbers add to the given total.

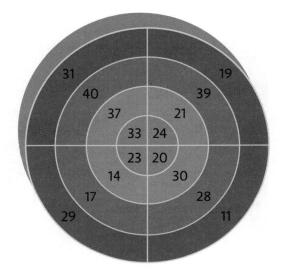

TOTALS: 77
         80
         126

SOLUTION page 120

62

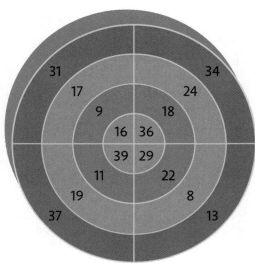

TOTALS: 61
         67
         117

SOLUTION page 122

LIGHTHOUSES

 Place a set of single-square ships in the grid. The number on each lighthouse reveals the total number of ships found in its row and column. Ships cannot touch either themselves or a lighthouse – not even diagonally. Ships must be in sight of a lighthouse, so may only be placed in a row or column which contains at least one lighthouse.

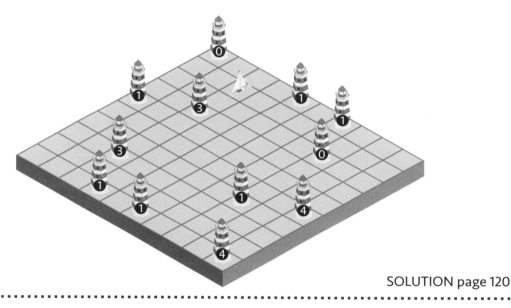

SOLUTION page 120

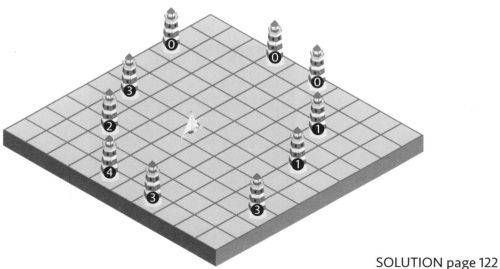

SOLUTION page 122

41

EASY AS A, B, C, D

65 Place a letter from A to D into some of the squares on the grid, so that each letter appears once in every row and column. This means that there will also be two empty squares in each row and column. Letters outside the grid reveal the first letter encountered in that row or column, from the viewpoint of the clue letter.

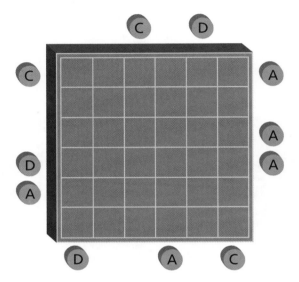

SOLUTION page 122

 66

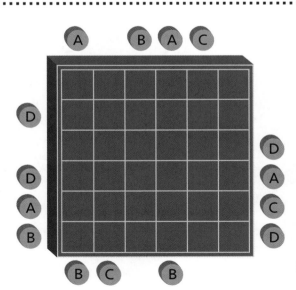

SOLUTION page 124

ARROW SUDOKU

67 Place numbers from 1 to 9 so that each number appears once in every row, column and 3x3 box. Also, each circled number must be equal to the sum of the numbers along its attached arrow.

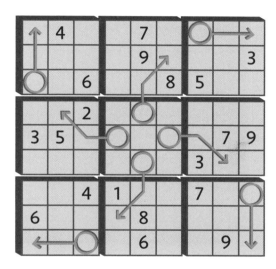

SOLUTION page 122

68

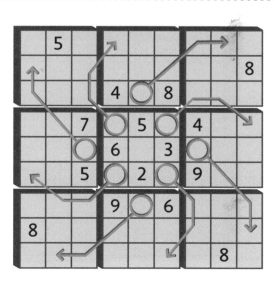

SOLUTION page 124

CLOUDS

 Shade some squares and/or rectangles of at least 2x2 in area, to form a set of clouds. Numbers outside the grid reveal the total number of shaded squares in each row and column. Clouds cannot touch – not even diagonally.

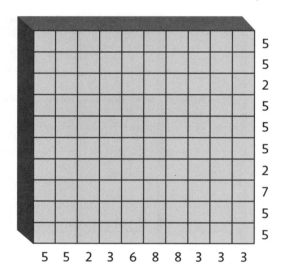

SOLUTION page 123

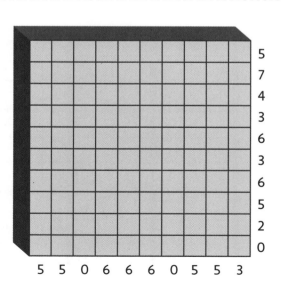

SOLUTION page 124

ARROWS

 Place an arrow in each box outside the grid so that each number inside the grid has the given number of arrows pointing at it. Arrows can point up, down, left, right or diagonally. All arrows must point to at least one number.

2	3	5	2
1	3	2	3
5	3	3	4
1	5	4	2

SOLUTION page 123

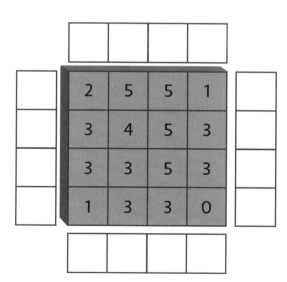

2	5	5	1
3	4	5	3
3	3	5	3
1	3	3	0

SOLUTION page 124

NO FOUR IN A ROW

 Place either an 'X' or an 'O' into every empty square, so that no lines of four or more 'X's or 'O's are made in any direction in the grid, including diagonally.

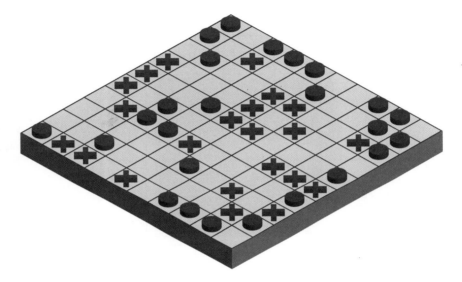

SOLUTION page 123

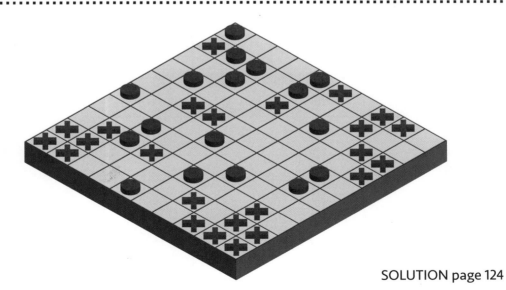

SOLUTION page 124

WRAPAROUND SUDOKU

 75 Place a letter from 1 to 9 into each empty square so that every letter appears once in each row, column and bold-lined jigsaw shape. Some jigsaw shapes 'wrap around' from one side of the grid to the other, continuing at the opposite end of the same row or column.

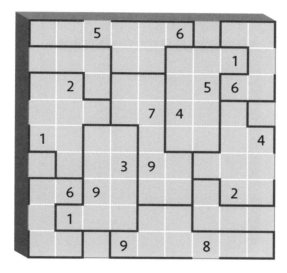

SOLUTION page 123

76

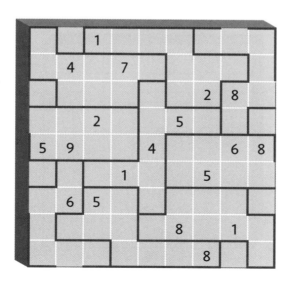

SOLUTION page 124

SKYSCRAPERS

77 Place a digit from 1 to 7 into every square, so that no digit repeats in any row or column inside the grid. Place digits in such a way that each given clue number outside the grid represents the number of digits that are 'visible' from that point, looking along that clue's row or column. A digit is visible if there is no higher digit preceding it.

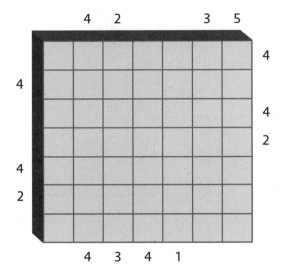

SOLUTION page 123

78

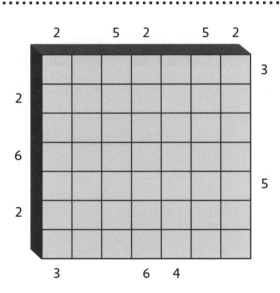

SOLUTION page 125

48

TRAIN TRACKS

79 Complete the railway track so that it enters and exits the grid only where shown. When the track enters a square, it can either turn 90 degrees or pass straight through the square. It cannot cross over itself. Numbers outside the grid reveal the total number of squares containing track in each row and column. Rows or columns without a number may contain any number of track segments.

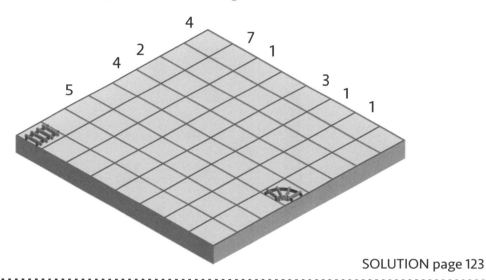

SOLUTION page 123

80

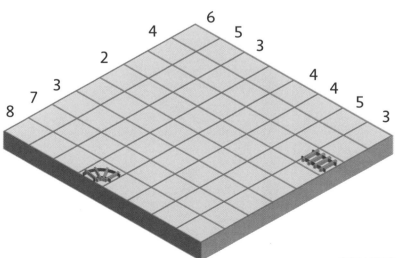

SOLUTION page 125

BLACKOUT SQUARE

Place a number from 1 to 8 into each square, so no number repeats within any row or column of the grid. Beware: each empty square may represent two different 'missing' numbers – one for the row, and one for the column.

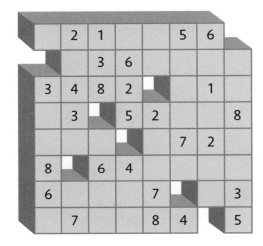

SOLUTION page 125

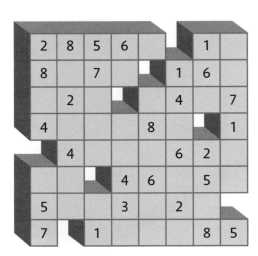

SOLUTION page 127

50

DOUBLE MINESWEEPER

83 Place one or two mines in some of the empty squares so that every given number has that many mines in its touching squares, including diagonally.

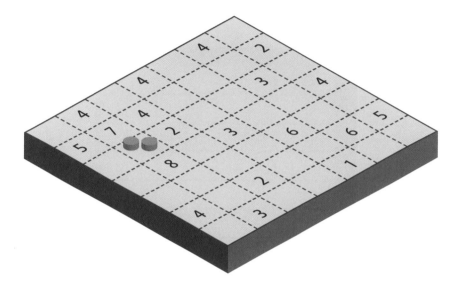

SOLUTION page 125

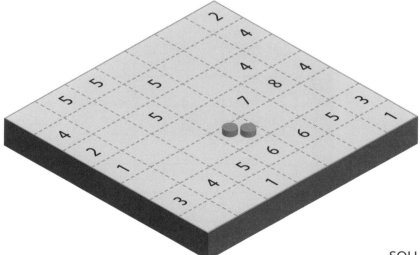

SOLUTION page 127

MYSTERY CALCUDOKU

85 Place a number from 1 to 8 once each into every row and column of the grid, while obeying the region clues. The value at the top left of each bold-lined region must be obtained when all of the numbers in that region have one of the four operations +, -, × and ÷ applied between them. To calculate - and ÷ results, begin with the largest number in the region and then subtract or divide by the other numbers in the region in any order.

2		75	2	21		9	
1960				8		12	
			12		120		21
5		20					
3	128					12	
			12		36		
4		2		18	19		
1		48				20	

SOLUTION page 125

86

21	2	96			1	90	
		42		19			6
14	11		14			21	
				280			240
	28					10	
24		16			2		
	48	15		420		4	7

SOLUTION page 127

LINESWEEPER

87 Draw a loop which travels horizontally and vertically through the centres of some of the empty squares. The loop must pass by each numbered square in the grid the given number of times. These counts represent the number of side touching and diagonally touching squares visited by the loop.

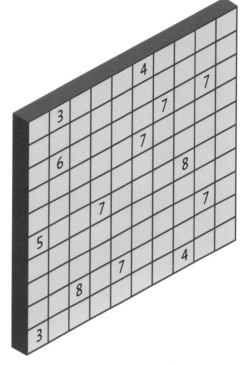

SOLUTION page 125

88

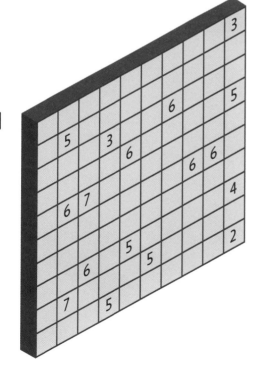

SOLUTION page 127

SHAPELINK DIAGONAL

89 Draw paths to join pairs of matching shapes. Paths can travel horizontally, vertically or at a 45-degree angle, and no more than one path can enter any square. Paths cannot cross, except diagonally on the intersection of two gridlines.

SOLUTION page 126

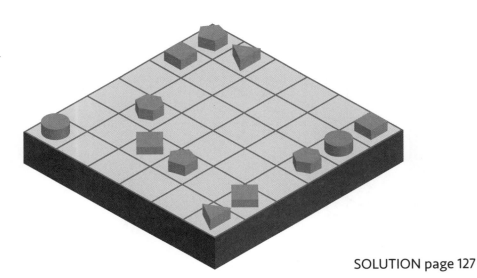

SOLUTION page 127

HASHI

91 Draw horizontal and vertical lines to join circled numbers. Each circle contains a number which specifies the number of lines that connect to it. No more than two lines may join any pair of circles. Lines may not cross other lines or circles. All circles must be joined in such a way that you can travel from any circle to any other circle by following one or more lines.

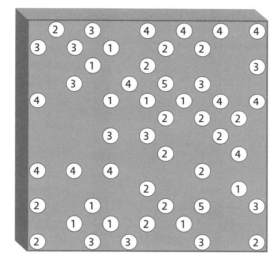

SOLUTION page 126

92

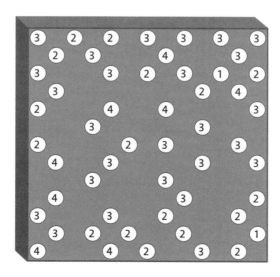

SOLUTION page 128

SPIRAL GALAXIES

93 Draw along some of the grid lines in order to divide the grid up into a set of regions. Every region must contain exactly one circle at its centre, and the region must be symmetrical in such a way that if rotated 180 degrees around the circle it would look exactly the same.

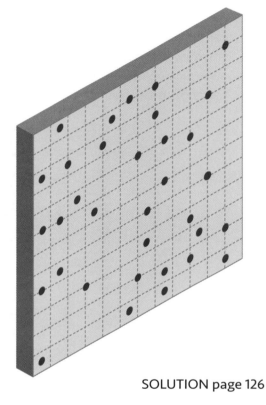

SOLUTION page 126

94

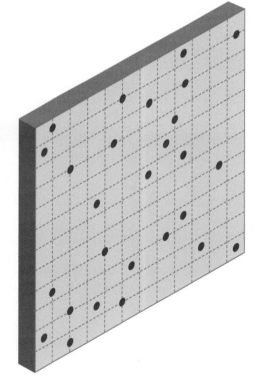

SOLUTION page 128

TOILS

 95 Shade squares on the grid in order to place all five of the tetrominoes (T, O, I, L and S as shown) once each into the grid, so that every numbered row or column contains that many shaded squares. Tetrominoes can be rotated or reflected, but may not touch one another – except diagonally.

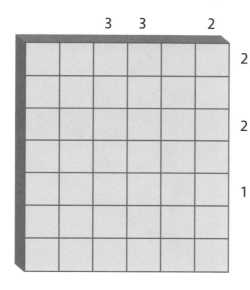

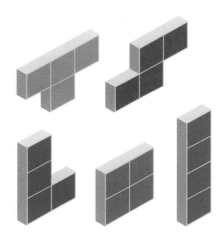

SOLUTION page 126

 96

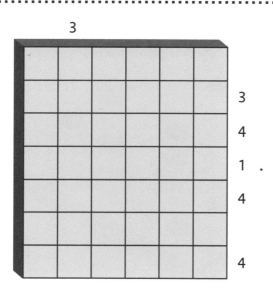

SOLUTION page 128

TENTS

97 Attach one tent to each tree, by placing it in a touching square in the same row or column. Numbers outside the grid reveal the total number of tents in certain rows and columns. Tents cannot touch each other – not even diagonally.

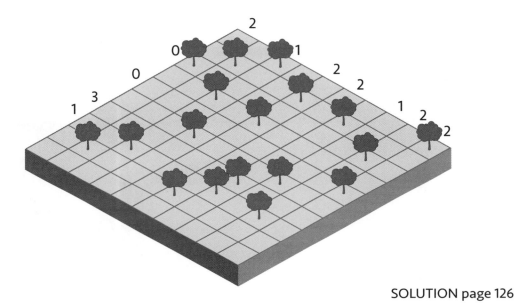

SOLUTION page 126

 98

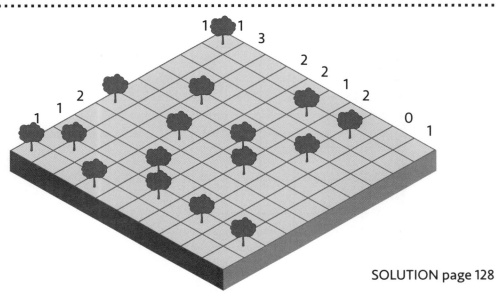

SOLUTION page 128

KING'S JOURNEY

99 Place a number into every empty square so that the numbers 1 to 100 appear once each in the grid. Also, a path from 1 to 100 must be formed which could be travelled step-by-step as a king moves in chess – i.e. starting at 1, the path must travel to 2, 3, 4, 5 and so on while moving only to side touching or diagonally touching squares.

		50		42	39		27		29
47	49				37			28	30
		44		20		24		31	33
1	2					22		34	
				67					15
	90					10	17		
87		92	93					63	
						58	64		62
	99		95		79				
100	97			83			76	73	

SOLUTION page 126

100

	57				39				
58		61				38	26	22	
48	50					42	29	27	
		51	45			36			
6						16	17		
	7			66	15				
	10		1		67	87	80		
				68					82
			92			72			75
98	96	100	94			70		74	

SOLUTION page 128

LIGHTHOUSES

101 Place a set of single-square ships in the grid. The number on each lighthouse reveals the total number of ships found in its row and column. Ships cannot touch either themselves or a lighthouse – not even diagonally. Ships must be in sight of a lighthouse, so may only be placed in a row or column which contains at least one lighthouse.

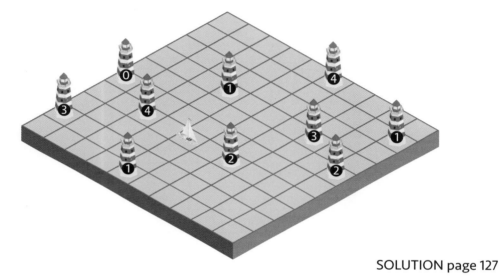

SOLUTION page 127

102

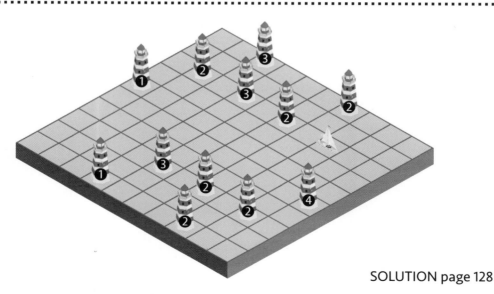

SOLUTION page 128

NUMBER DARTS

 Can you make each of the totals shown? For each total, choose one number from each of the four rings, so that those four numbers add to the given total.

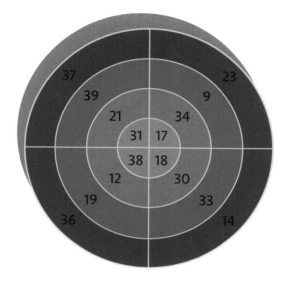

TOTALS: 77
79
129

SOLUTION page 129

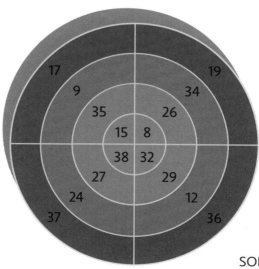

TOTALS: 77
100
131

SOLUTION page 132

EASY AS A, B, C, D

105 Place a letter from A to D into some of the squares on the grid, so that each letter appears once in every row and column. This means that there will also be two empty squares in each row and column. Letters outside the grid reveal the first letter encountered in that row or column, from the viewpoint of the clue letter.

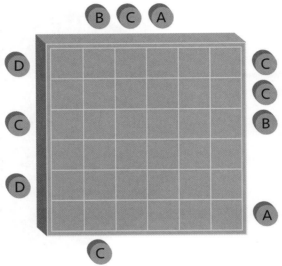

SOLUTION page 129

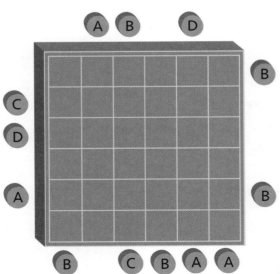

SOLUTION page 132

ARROW SUDOKU

 107 Place numbers from 1 to 9 so that each number appears once in every row, column and 3x3 box. Also, each circled number must be equal to the sum of the numbers along its attached arrow.

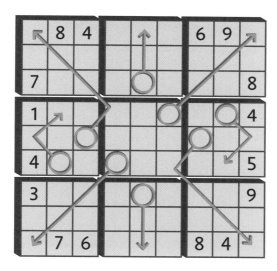

SOLUTION page 129

 108

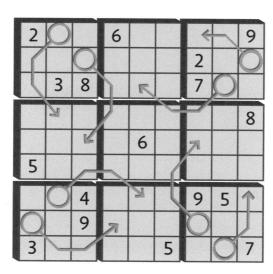

SOLUTION page 132

CLOUDS

109 Shade some squares and/or rectangles of at least 2x2 in area, to form a set of clouds. Numbers outside the grid reveal the total number of shaded squares in each row and column. Clouds cannot touch – not even diagonally.

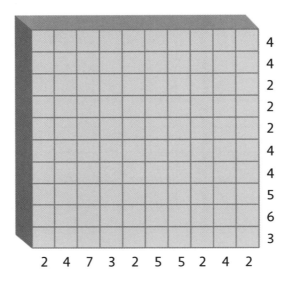

4
4
2
2
2
4
4
5
6
3

2 4 7 3 2 5 5 2 4 2

SOLUTION page 129

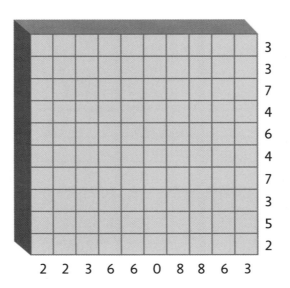

3
3
7
4
6
4
7
3
5
2

2 2 3 6 6 0 8 8 6 3

SOLUTION page 132

ARROWS

 Place an arrow in each box outside the grid so that each number inside the grid has the given number of arrows pointing at it. Arrows can point up, down, left, right or diagonally. All arrows must point to at least one number.

1	4	3	4
3	2	3	5
2	1	1	4
2	2	1	3

SOLUTION page 129

4	5	4	3
3	2	1	1
5	4	2	3
3	4	3	2

SOLUTION page 133

NO FOUR IN A ROW

113 Place either an 'X' or an 'O' into every empty square, so that no lines of four or more 'X's or 'O's are made in any direction in the grid, including diagonally.

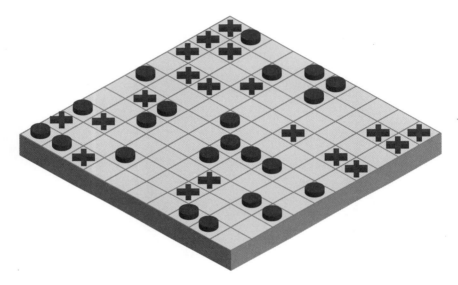

SOLUTION page 129

114

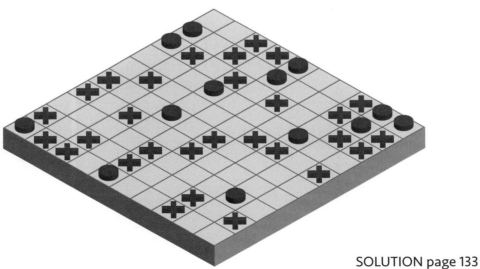

SOLUTION page 133

WRAPAROUND SUDOKU

 Place a letter from 1 to 9 into each empty square so that every letter appears once in each row, column and bold-lined jigsaw shape. Some jigsaw shapes 'wrap around' from one side of the grid to the other, continuing at the opposite end of the same row or column.

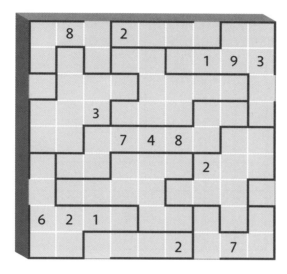

SOLUTION page 130

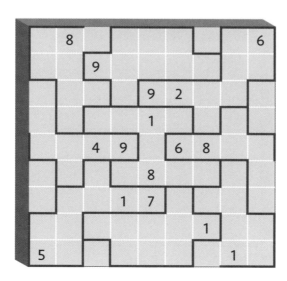

SOLUTION page 133

SKYSCRAPERS

117 Place a digit from 1 to 7 into every square, so that no digit repeats in any row or column inside the grid. Place digits in such a way that each given clue number outside the grid represents the number of digits that are 'visible' from that point, looking along that clue's row or column. A digit is visible if there is no higher digit preceding it.

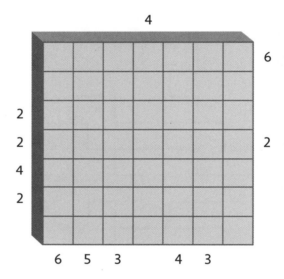

SOLUTION page 133

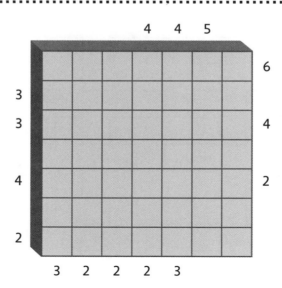

SOLUTION page 130

TRAIN TRACKS

119 Complete the railway track so that it enters and exits the grid only where shown. When the track enters a square, it can either turn 90 degrees or pass straight through the square. It cannot cross over itself. Numbers outside the grid reveal the total number of squares containing track in each row and column. Rows or columns without a number may contain any number of track segments.

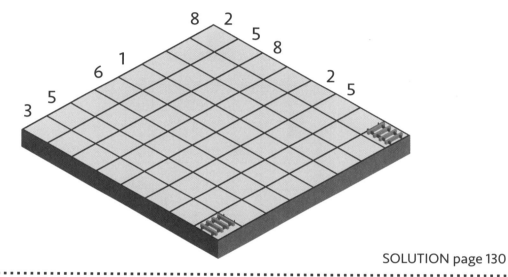

SOLUTION page 130

120

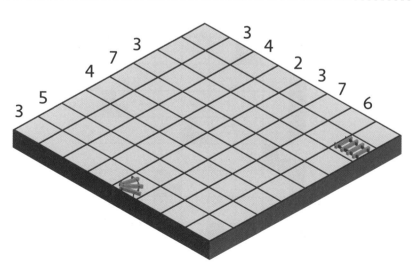

SOLUTION page 133

BLACKOUT SQUARE

121 Place a number from 1 to 8 into each square, so no number repeats within any row or column of the grid. Beware: each empty square may represent two different 'missing' numbers – one for the row, and one for the column.

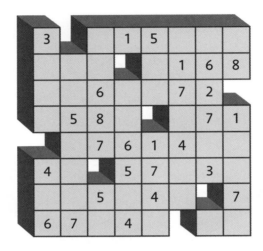

SOLUTION page 130

122

SOLUTION page 133

DOUBLE MINESWEEPER

 Place one or two mines in some of the empty squares so that every given number has that many mines in its touching squares, including diagonally.

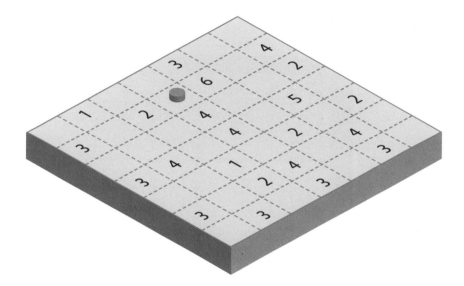

SOLUTION page 130

 124

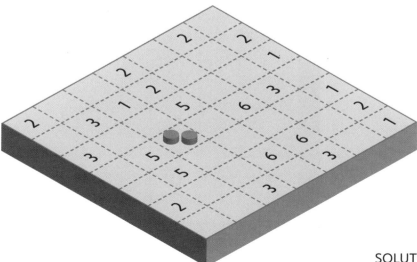

SOLUTION page 134

MYSTERY CALCUDOKU

125 Place a number from 1 to 8 once each into every row and column of the grid, while obeying the region clues. The value at the top left of each bold-lined region must be obtained when all of the numbers in that region have one of the four operations +, -, × and ÷ applied between them. To calculate - and ÷ results, begin with the largest number in the region and then subtract or divide by the other numbers in the region in any order.

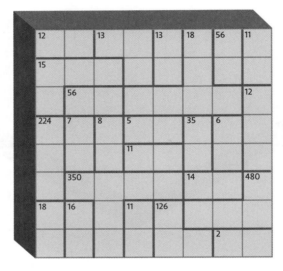

SOLUTION page 130

SOLUTION page 134

LINESWEEPER

127 Draw a loop which travels horizontally and vertically through the centres of some of the empty squares. The loop must pass by each numbered square in the grid the given number of times. These counts represent the number of side touching and diagonally touching squares visited by the loop.

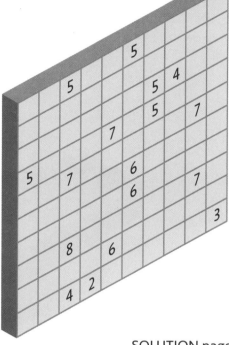

SOLUTION page 131

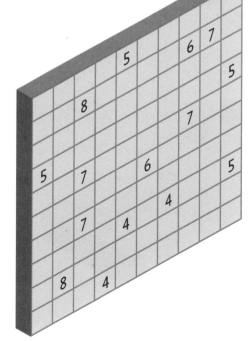

SOLUTION page 134

SHAPELINK DIAGONAL

129 Draw paths to join pairs of matching shapes. Paths can travel horizontally, vertically or at a 45-degree angle, and no more than one path can enter any square. Paths cannot cross, except diagonally on the intersection of two gridlines.

SOLUTION page 131

SOLUTION page 134

HASHI

131 Draw horizontal and vertical lines to join circled numbers. Each circle contains a number which specifies the number of lines that connect to it. No more than two lines may join any pair of circles. Lines may not cross other lines or circles. All circles must be joined in such a way that you can travel from any circle to any other circle by following one or more lines.

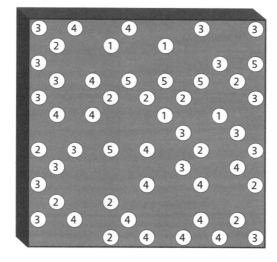

SOLUTION page 131

132

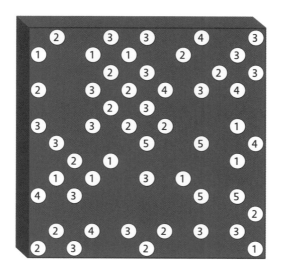

SOLUTION page 134

SPIRAL GALAXIES

133 Draw along some of the grid lines in order to divide the grid up into a set of regions. Every region must contain exactly one circle at its centre, and the region must be symmetrical in such a way that if rotated 180 degrees around the circle it would look exactly the same.

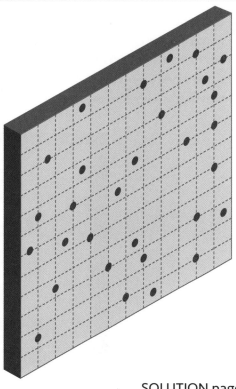

SOLUTION page 131

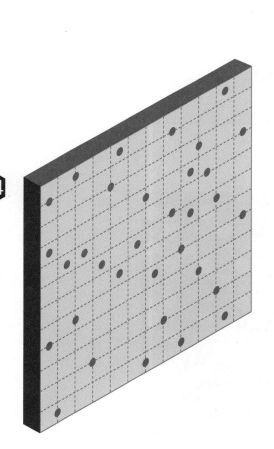

134

SOLUTION page 134

TOILS

135 Shade squares on the grid in order to place all five of the tetrominoes (T, O, I, L and S as shown) once each into the grid, so that every numbered row or column contains that many shaded squares. Tetrominoes can be rotated or reflected, but may not touch one another – except diagonally.

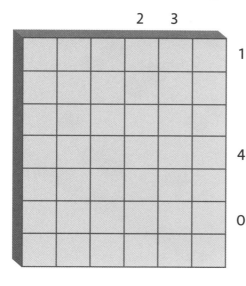

SOLUTION page 131

136

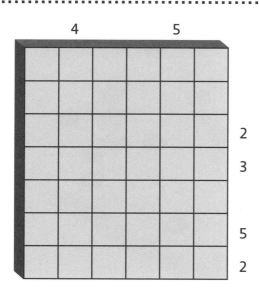

SOLUTION page 135

TENTS

137 Attach one tent to each tree, by placing it in a touching square in the same row or column. Numbers outside the grid reveal the total number of tents in certain rows and columns. Tents cannot touch each other – not even diagonally.

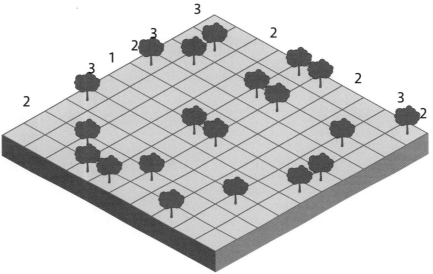

SOLUTION page 131

138

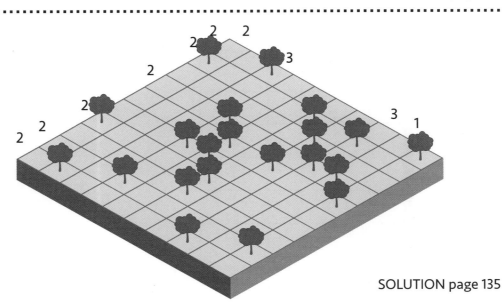

SOLUTION page 135

KING'S JOURNEY

 Place a number into every empty square so that the numbers 1 to 100 appear once each in the grid. Also, a path from 1 to 100 must be formed which could be travelled step-by-step as a king moves in chess – i.e. starting at 1, the path must travel to 2, 3, 4, 5 and so on while moving only to side touching or diagonally touching squares.

139

	8	6					1		67
	9		81		72			66	65
12		82		80		71	61		
	14	19			77			60	57
	18		85			59			
		94	91			48	50	55	
		97		90	88				
23						42			
	27	30	31	100	34	41			
	28			33		36			

SOLUTION page 132

140

				36			26	100	99
44		40			28	24			98
	46		34		23				97
50	51		32		22			94	91
			73	74	17			89	
				76	81				86
70	71		79						
68		56							6
67	63			58		2	3	4	
		62	61	59	1				8

SOLUTION page 135

LIGHTHOUSES

141 Place a set of single-square ships in the grid. The number on each lighthouse reveals the total number of ships found in its row and column. Ships cannot touch either themselves or a lighthouse – not even diagonally. Ships must be in sight of a lighthouse, so may only be placed in a row or column which contains at least one lighthouse.

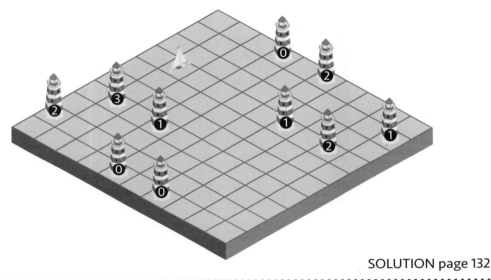

SOLUTION page 132

142

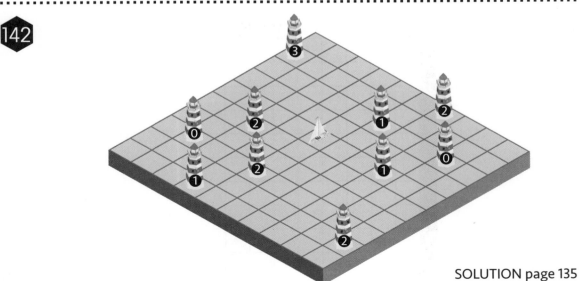

SOLUTION page 135

NUMBER DARTS

 143 Can you make each of the totals shown? For each total, choose one number from each of the four rings, so that those four numbers add to the given total.

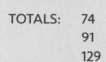

TOTALS: 74
         91
         129

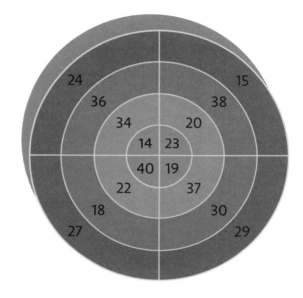

SOLUTION page 135

 144

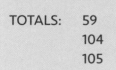

TOTALS: 59
         104
         105

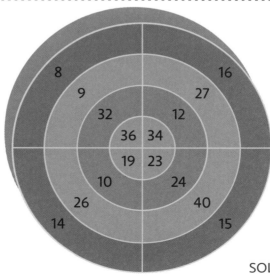

SOLUTION page 137

EASY AS A, B, C, D

145 Place a letter from A to D into some of the squares on the grid, so that each letter appears once in every row and column. This means that there will also be two empty squares in each row and column. Letters outside the grid reveal the first letter encountered in that row or column, from the viewpoint of the clue letter.

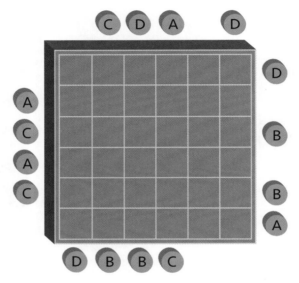

SOLUTION page 135

 146

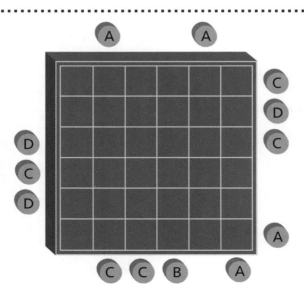

SOLUTION page 137

ARROW SUDOKU

147 Place numbers from 1 to 9 so that each number appears once in every row, column and 3x3 box. Also, each circled number must be equal to the sum of the numbers along its attached arrow.

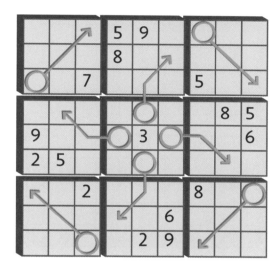

SOLUTION page 136

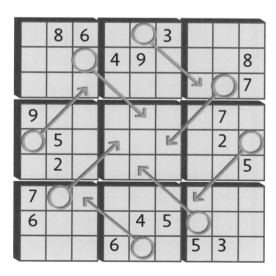

SOLUTION page 137

CLOUDS

149 Shade some squares and/or rectangles of at least 2x2 in area, to form a set of clouds. Numbers outside the grid reveal the total number of shaded squares in each row and column. Clouds cannot touch – not even diagonally.

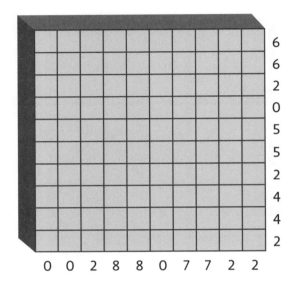

Row clues (top to bottom): 6, 6, 2, 0, 5, 5, 2, 4, 4, 2

Column clues (left to right): 0, 0, 2, 8, 8, 0, 7, 7, 2, 2

SOLUTION page 136

150

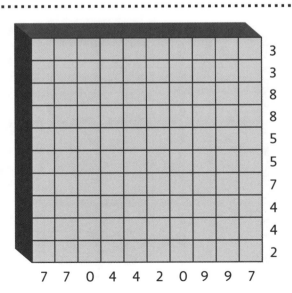

Row clues (top to bottom): 3, 3, 8, 8, 5, 5, 7, 4, 4, 2

Column clues (left to right): 7, 7, 0, 4, 4, 2, 0, 9, 9, 7

SOLUTION page 137

ARROWS

151 Place an arrow in each box outside the grid so that each
number inside the grid has the given number of arrows
pointing at it. Arrows can point up, down, left, right or
diagonally. All arrows must point to at least one number.

4	3	4	5
3	1	1	2
5	2	1	3
4	1	1	2

SOLUTION page 136

152

2	3	4	2
3	2	4	4
2	3	4	5
3	4	7	3

SOLUTION page 137

NO FOUR IN A ROW

 153 Place either an 'X' or an 'O' into every empty square, so that no lines of four or more 'X's or 'O's are made in any direction in the grid, including diagonally.

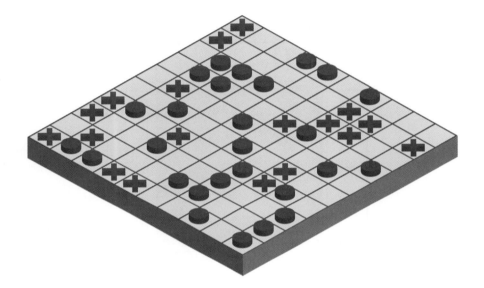

SOLUTION page 136

154

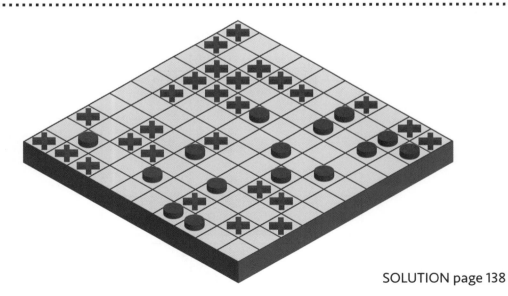

SOLUTION page 138

WRAPAROUND SUDOKU

155 Place a letter from 1 to 9 into each empty square so that every letter
appears once in each row, column and bold-lined jigsaw shape.
Some jigsaw shapes 'wrap around' from one side of the grid to the
other, continuing at the opposite end of the same row or column.

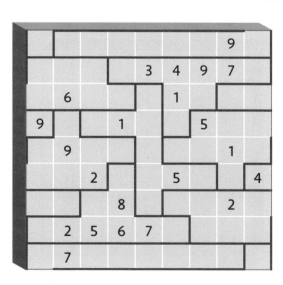

SOLUTION page 136

SOLUTION page 138

87

SKYSCRAPERS

157 Place a digit from 1 to 7 into every square, so that no digit repeats in any row or column inside the grid. Place digits in such a way that each given clue number outside the grid represents the number of digits that are 'visible' from that point, looking along that clue's row or column. A digit is visible if there is no higher digit preceding it.

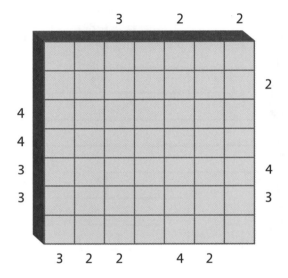

SOLUTION page 136

158

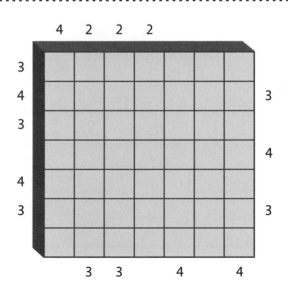

SOLUTION page 138

TRAIN TRACKS

 159 Complete the railway track so that it enters and exits the grid only where shown. When the track enters a square, it can either turn 90 degrees or pass straight through the square. It cannot cross over itself. Numbers outside the grid reveal the total number of squares containing track in each row and column. Rows or columns without a number may contain any number of track segments.

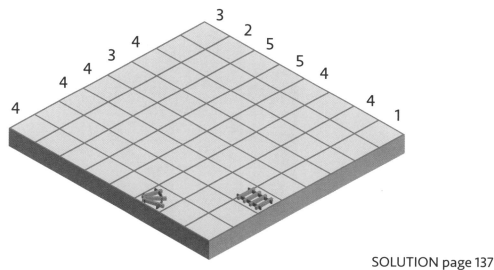

SOLUTION page 137

 160

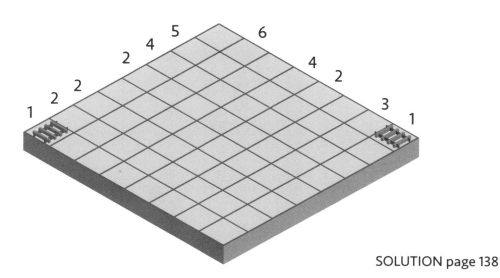

SOLUTION page 138

BLACKOUT SQUARE

161 Place a number from 1 to 8 into each square, so no number repeats within any row or column of the grid. Beware: each empty square may represent two different 'missing' numbers – one for the row, and one for the column.

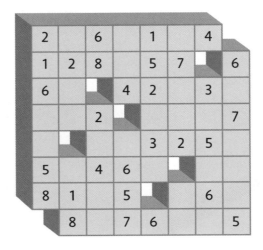

SOLUTION page 138

SOLUTION page 140

DOUBLE MINESWEEPER

 163 Place one or two mines in some of the empty squares so that every given number has that many mines in its touching squares, including diagonally.

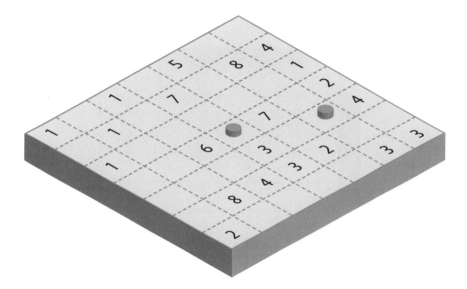

SOLUTION page 140

 164

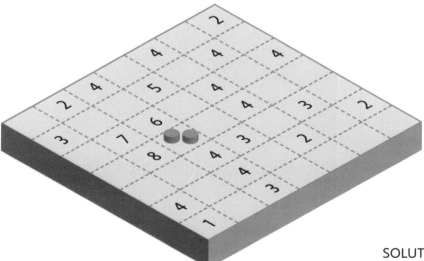

SOLUTION page 138

MYSTERY CALCUDOKU

 165 Place a number from 1 to 8 once each into every row and column of the grid, while obeying the region clues. The value at the top left of each bold-lined region must be obtained when all of the numbers in that region have one of the four operations +, -, × and ÷ applied between them. To calculate - and ÷ results, begin with the largest number in the region and then subtract or divide by the other numbers in the region in any order.

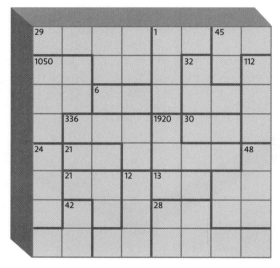

SOLUTION page 140

166

SOLUTION page 139

LINESWEEPER

167 Draw a loop which travels horizontally and vertically through the centres of some of the empty squares. The loop must pass by each numbered square in the grid the given number of times. These counts represent the number of side touching and diagonally touching squares visited by the loop.

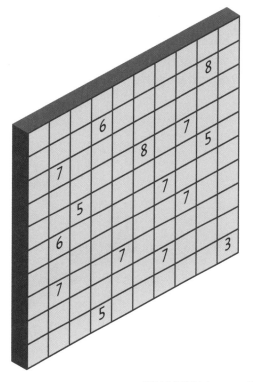

SOLUTION page 139

168

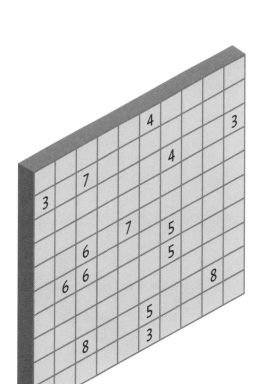

SOLUTION page 141

SHAPELINK DIAGONAL

169 Draw paths to join pairs of matching shapes. Paths can travel horizontally, vertically or at a 45-degree angle, and no more than one path can enter any square. Paths cannot cross, except diagonally on the intersection of two gridlines.

SOLUTION page 139

SOLUTION page 141

HASHI

 Draw horizontal and vertical lines to join circled numbers. Each circle contains a number which specifies the number of lines that connect to it. No more than two lines may join any pair of circles. Lines may not cross other lines or circles. All circles must be joined in such a way that you can travel from any circle to any other circle by following one or more lines.

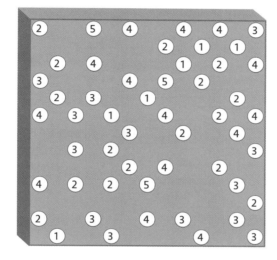

SOLUTION page 139

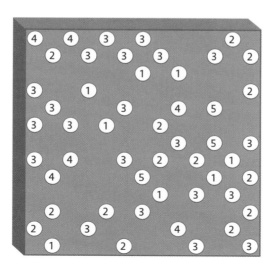

SOLUTION page 141

SPIRAL GALAXIES

173 Draw along some of the grid lines in order to divide the grid up into a set of regions. Every region must contain exactly one circle at its centre, and the region must be symmetrical in such a way that if rotated 180 degrees around the circle it would look exactly the same.

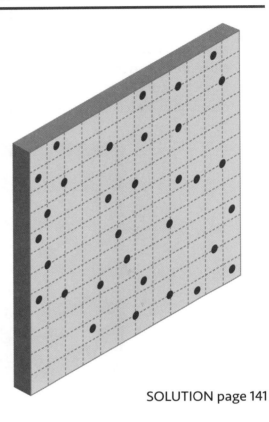

SOLUTION page 141

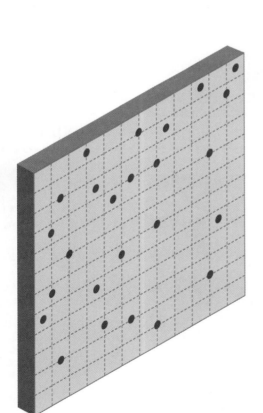

174

SOLUTION page 139

TOILS

175 Shade squares on the grid in order to place all five of the
tetrominoes (T, O, I, L and S as shown) once each into the grid,
so that every numbered row or column contains that many
shaded squares. Tetrominoes can be rotated or reflected, but
may not touch one another – except diagonally.

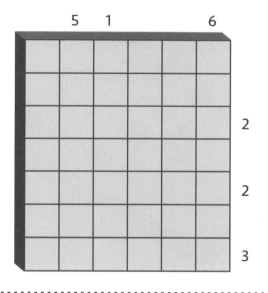

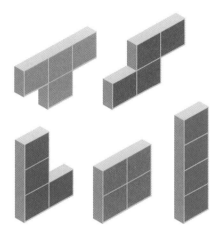

SOLUTION page 139

176

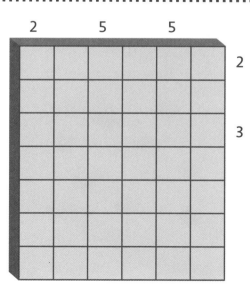

SOLUTION page 141

TENTS

177 Attach one tent to each tree, by placing it in a touching square in the same row or column. Numbers outside the grid reveal the total number of tents in certain rows and columns. Tents cannot touch each other – not even diagonally.

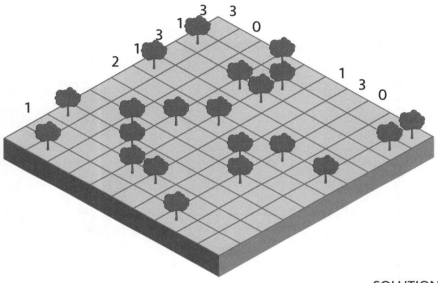

SOLUTION page 140

178

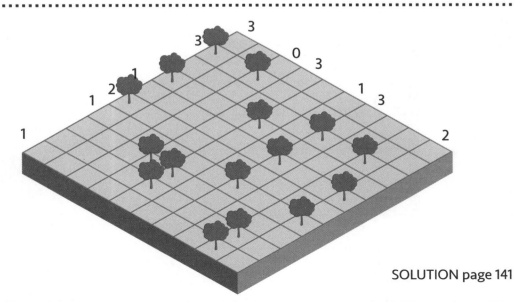

SOLUTION page 141

KING'S JOURNEY

179 Place a number into every empty square so that the numbers 1 to 100 appear once each in the grid. Also, a path from 1 to 100 must be formed which could be travelled step-by-step as a king moves in chess – i.e. starting at 1, the path must travel to 2, 3, 4, 5 and so on while moving only to side touching or diagonally touching squares.

		16		12			43		
	27		15					42	
	26	32	21		9	8		49	
							50		39
			34		36				
90									56
89			98		100		61	57	
92				1		3			58
94		75		70		63	64		
		73							66

SOLUTION page 140

180

	98	87	86	85	83		18		
100			89		82			19	
95			91					1	
					12			32	
	73	79	39			11		33	31
	75				36				30
		77		41		25			29
	70		42	44					
		68	60		57		47		49
64		62		58	56	54		52	50

SOLUTION page 142

LIGHTHOUSES

181 Place a set of single-square ships in the grid. The number on each lighthouse reveals the total number of ships found in its row and column. Ships cannot touch either themselves or a lighthouse – not even diagonally. Ships must be in sight of a lighthouse, so may only be placed in a row or column which contains at least one lighthouse.

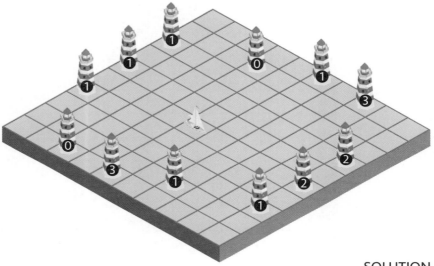

SOLUTION page 140

182

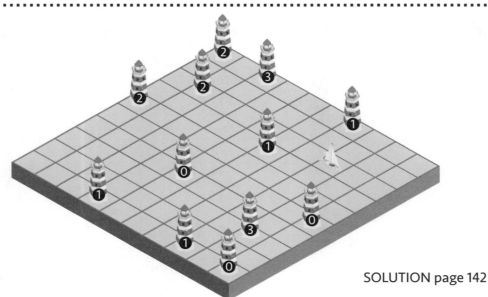

SOLUTION page 142

NUMBER DARTS

183 Can you make each of the totals shown? For each total, choose one number from each of the four rings, so that those four numbers add to the given total.

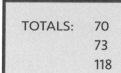

TOTALS: 70
73
118

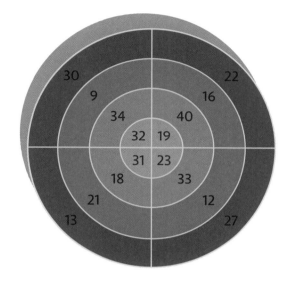

SOLUTION page 140

184

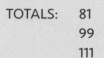

TOTALS: 81
99
111

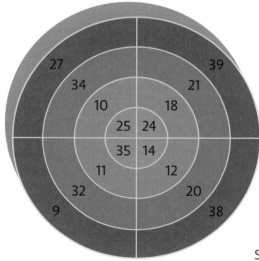

SOLUTION page 142

EASY AS A, B, C, D

185 Place a letter from A to D into some of the squares on the grid, so that each letter appears once in every row and column. This means that there will also be two empty squares in each row and column. Letters outside the grid reveal the first letter encountered in that row or column, from the viewpoint of the clue letter.

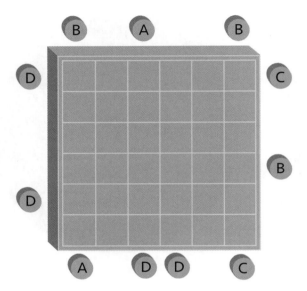

SOLUTION page 142

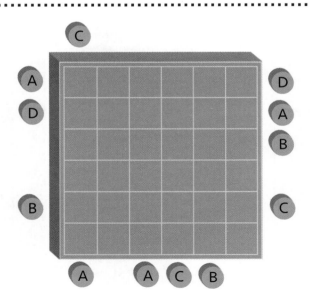

SOLUTION page 143

ARROW SUDOKU

187 Place numbers from 1 to 9 so that each number appears once in every row, column and 3x3 box. Also, each circled number must be equal to the sum of the numbers along its attached arrow.

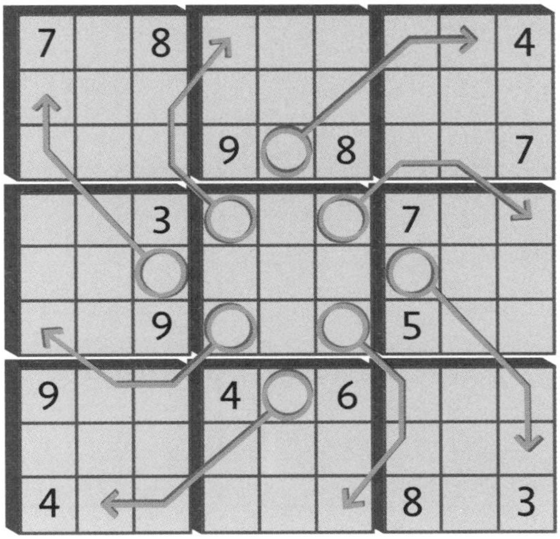

SOLUTION page 142

188

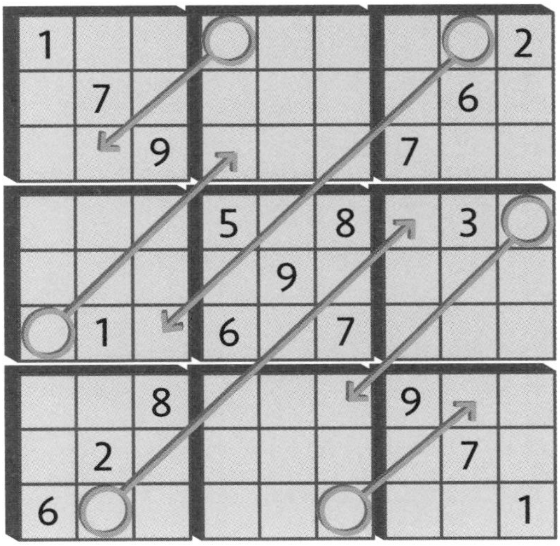

SOLUTION page 144

CLOUDS

189 Shade some squares and/or rectangles of at least 2x2 in area, to form a set of clouds. Numbers outside the grid reveal the total number of shaded squares in each row and column. Clouds cannot touch – not even diagonally.

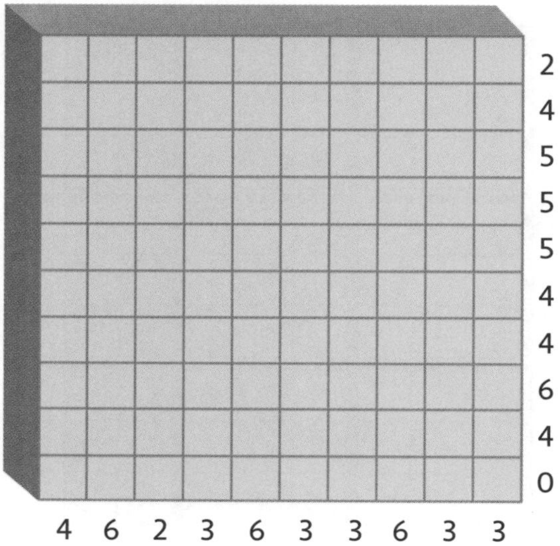
2
4
5
5
5
4
4
6
4
0
4 6 2 3 6 3 3 6 3 3

SOLUTION page 142

190

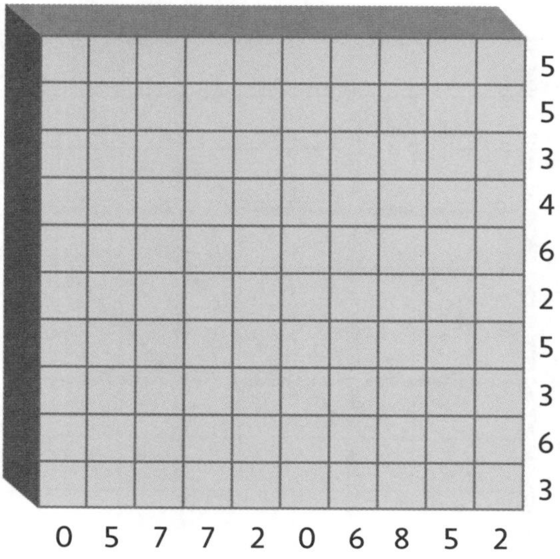
5
5
3
4
6
2
5
3
6
3
0 5 7 7 2 0 6 8 5 2

SOLUTION page 144

ARROWS

191 Place an arrow in each box outside the grid so that each number inside the grid has the given number of arrows pointing at it. Arrows can point up, down, left, right or diagonally. All arrows must point to at least one number.

3	2	4	3
1	2	2	2
6	1	3	4
3	4	3	2

SOLUTION page 143

4	3	2	5
4	1	1	3
2	2	2	3
3	3	4	4

SOLUTION page 144

NO FOUR IN A ROW

193 Place either an 'X' or an 'O' into every empty square, so that no lines of four or more 'X's or 'O's are made in any direction in the grid, including diagonally.

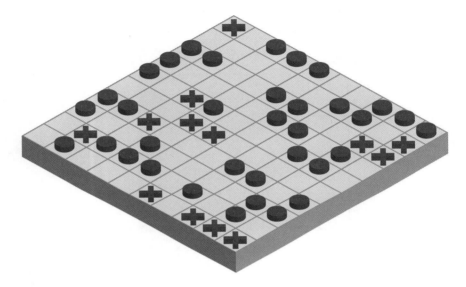

SOLUTION page 143

194

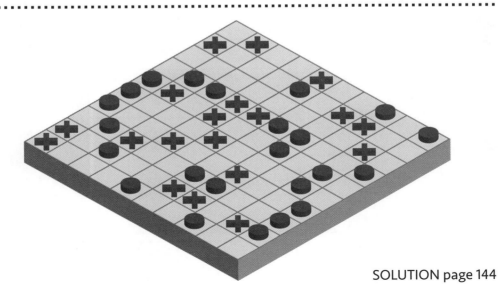

SOLUTION page 144

WRAPAROUND SUDOKU

195 Place a letter from 1 to 9 into each empty square so that every letter appears once in each row, column and bold-lined jigsaw shape. Some jigsaw shapes 'wrap around' from one side of the grid to the other, continuing at the opposite end of the same row or column.

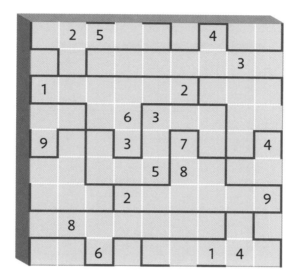

SOLUTION page 143

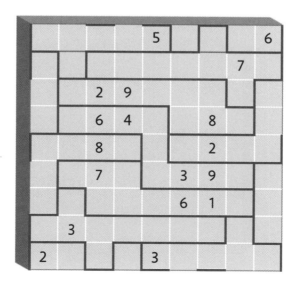

SOLUTION page 144

SKYSCRAPERS

197 Place a digit from 1 to 7 into every square, so that no digit repeats in any row or column inside the grid. Place digits in such a way that each given clue number outside the grid represents the number of digits that are 'visible' from that point, looking along that clue's row or column. A digit is visible if there is no higher digit preceding it.

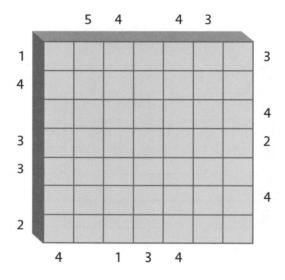

SOLUTION page 144

198

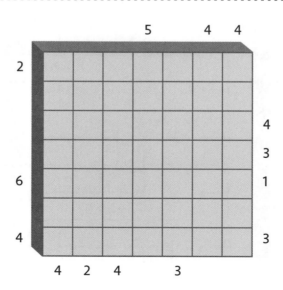

SOLUTION page 143

TRAIN TRACKS

 Complete the railway track so that it enters and exits the grid only where shown. When the track enters a square, it can either turn 90 degrees or pass straight through the square. It cannot cross over itself. Numbers outside the grid reveal the total number of squares containing track in each row and column. Rows or columns without a number may contain any number of track segments.

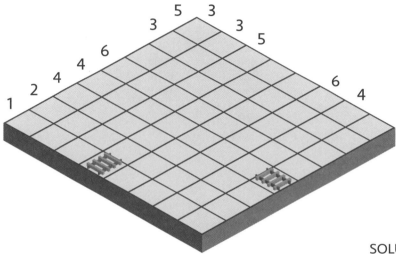

SOLUTION page 143

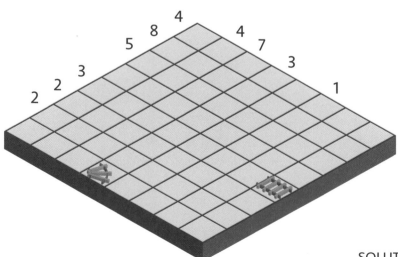

SOLUTION page 142

SOLUTIONS

SOLUTIONS

01

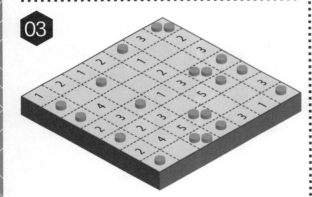

1	7		8	6	3	2	4
2		6	7	5	1	4	3
5	2	7	6		4	1	8
6	4	2	3	8	5		1
4	5	8	1	4	2	3	
4	3	1		2	7	6	5
	8	3	4	1	6	5	2
8	1	4	2	3		7	6

03

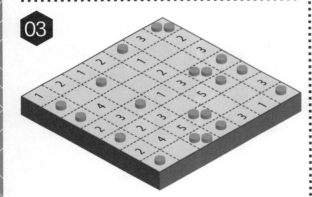

05

^{13}8	5	6	^{10}7	3	^{6}2	^{0}1	4
^{8}1	8	2	^{90}5	^{42}6	4	^{245}7	3
^{20}4	^{2}2	3	6	1	^{120}8	5	7
5	4	^{19}8	1	7	3	^{3}2	6
^{14}2	7	4	^{9}3	^{480}8	5	^{48}6	^{6}1
^{1}3	1	7	4	2	6	8	5
^{252}6	3	^{6}1	2	5	^{6}7	^{4}4	8
7	6	5	^{32}8	4	1	^{6}3	2

07

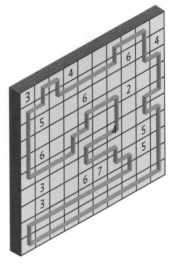

09

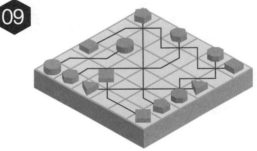

11

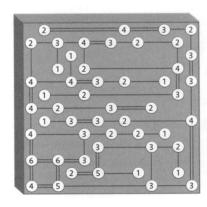

13

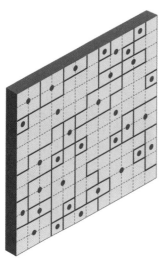

19

37	36	20	21	18	15	14	12	9	10
38	39	35	19	22	17	16	13	11	8
40	41	34	31	29	23	24	25	7	1
42	46	33	32	30	28	26	100	6	2
43	45	47	48	49	27	99	97	5	3
44	52	51	50	59	60	98	96	95	4
53	56	57	58	61	62	93	94	86	85
54	55	65	64	63	92	91	88	87	84
69	70	66	74	73	90	89	79	83	81
68	67	71	72	75	76	77	78	80	82

21

78 = 10 + 19 + 36 + 13
95 = 20 + 8 + 27 + 40
128 = 39 + 19 + 36 + 34

15

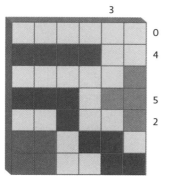

23

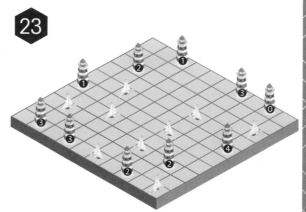

17

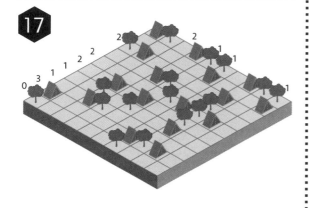

SOLUTIONS

02

2		8	4	7	1	5	6
4	7	2	5	1		8	3
5	4	3		2	7	1	8
	3	7	2	5	4	6	1
1	5	4	3		8	2	7
8	2	5	1	4	6	3	
6	1		8	3	5	4	2
3	8	1	7	6	2		5

04

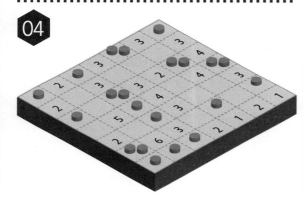

06

08

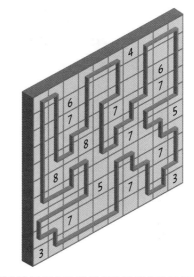

10

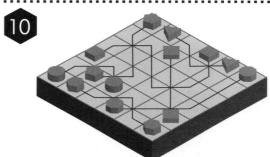

12

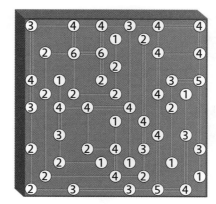

14

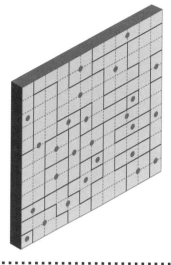

16

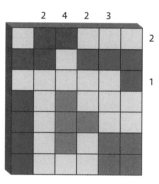

18

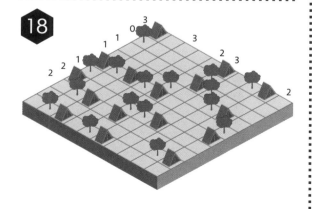

20

61	63	59	58	46	45	22	20	19	17
62	60	64	57	47	44	23	21	16	18
69	67	65	56	43	48	24	15	11	13
70	68	66	55	49	42	25	10	14	12
73	71	54	50	41	26	9	6	7	34
72	74	53	51	40	27	5	8	35	33
79	77	75	52	39	4	28	36	30	32
78	80	76	2	3	38	37	29	100	31
84	82	81	1	88	90	92	99	95	97
83	85	86	87	89	91	93	94	98	96

22

64 = 18 + 19 + 16 + 11

71 = 10 + 21 + 16 + 24

125 = 20 + 37 + 35 + 33

24

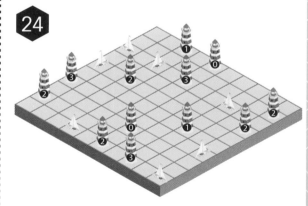

115

SOLUTIONS

25

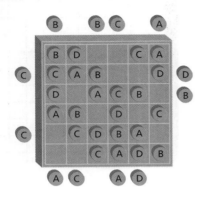

31

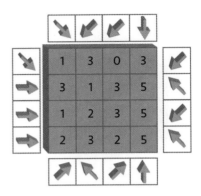

27

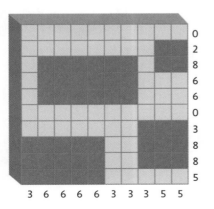

33

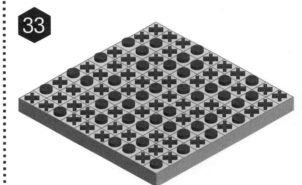

29

										0
										2
										8
										6
										6
										0
										3
										8
										8
										5

3 6 6 6 6 3 3 3 5 5

35

1	5	3	9	6	2	8	7	4
3	8	2	4	1	6	9	5	7
4	9	7	2	3	8	6	1	5
8	3	6	5	4	1	7	9	2
9	1	4	7	8	5	3	2	6
7	6	5	3	2	9	1	4	8
6	2	1	8	7	4	5	3	9
5	4	8	1	9	7	2	6	3
2	7	9	6	5	3	4	8	1

37

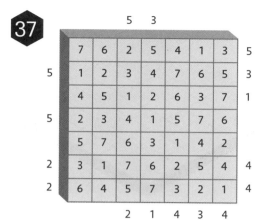

39

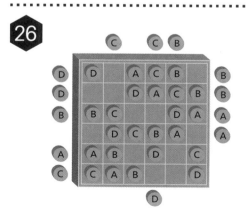

26

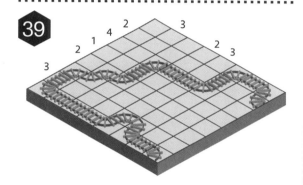

28

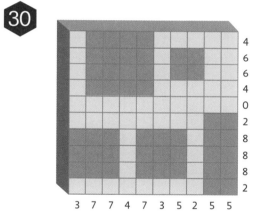

30

32

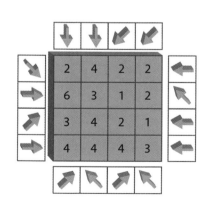

117

SOLUTIONS

34

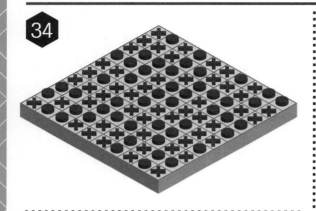

40

36

2	8	3	1	9	4	5	7	6
6	2	5	7	3	9	8	4	1
8	3	9	2	4	5	6	1	7
5	1	2	9	8	7	4	6	3
3	6	7	5	2	8	1	9	4
7	5	8	4	1	6	2	3	9
1	4	6	3	7	2	9	5	8
4	9	1	8	6	3	7	2	5
9	7	4	6	5	1	3	8	2

41

5	1	6	2	8		3	4
	8	2	3	6	7	4	1
3	4		1	5	2	8	7
8	5	4	7		1	2	3
7	2	1	6	3	5		8
1	3	5		4	8	7	6
6		3	8	7	4	1	2
2	6	7	4	1	3	5	

38

	4	5				3		
3	7	3	1	6	2	5	4	**4**
	2	5	4	7	6	1	3	**3**
4	4	1	5	2	3	7	6	**2**
5	3	4	6	5	1	2	7	
	1	2	3	4	7	6	5	**3**
	5	6	7	1	4	3	2	**4**
	6	7	2	3	5	4	1	
				4	2		4	

43

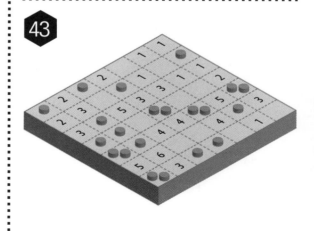

45

[11] 2	3	[56] 7	8	4	5	[1] 1	[6] 6
6	[21] 7	3	[24] 4	[56] 8	1	[9] 5	2
[0] 1	[48] 6	8	3	[12] 5	7	2	[13] 4
8	[20] 5	4	2	7	[9] 6	3	1
4	[10] 2	5	[6] 1	[18] 6	[56] 8	7	3
3	[160] 8	[12] 2	7	1	[24] 4	6	5
5	4	1	6	3	[16] 2	8	[224] 7
[7] 7	1	[30] 6	5	[1] 2	3	4	8

51

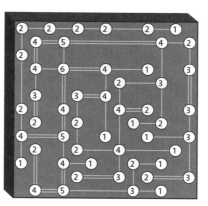

47

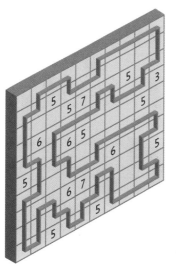

53

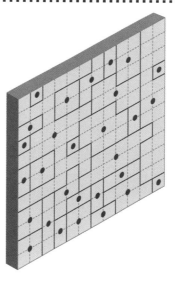

49

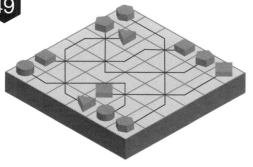

55

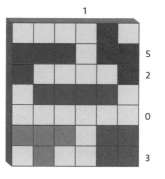

119

SOLUTIONS

57

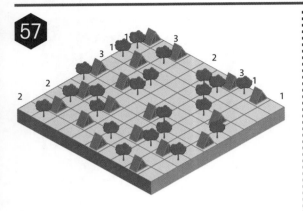

63

59

77	79	75	74	73	83	84	68	69	1
78	76	80	81	82	72	85	70	67	2
94	95	89	97	87	86	71	66	3	5
93	90	96	88	98	99	65	25	6	4
91	92	61	62	100	64	26	24	23	7
57	58	59	60	63	31	27	22	10	8
56	54	53	52	32	28	30	21	11	9
55	49	43	51	41	33	29	20	13	12
48	44	50	42	40	35	34	19	16	14
47	46	45	39	38	37	36	18	17	15

42

3	6	8		4	1	5	2
8	1		7	2	3	4	6
4	8	6	2		7	1	5
6	2	7	4	1		8	3
2	4	1	5	3	6	7	
	7	2	1	5	8	3	4
7	3	4	6	8	2		1
1		5	3	6	4	2	8

61

77 = 24 + 14 + 28 + 11

80 = 23 + 21 + 17 + 19

126 = 36 + 21 + 40 + 29

44

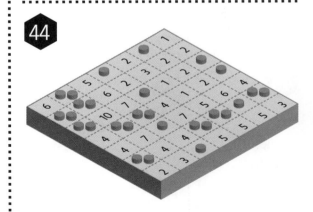

46

21 3	32 8	2 5	2	1	7 7	2 4	6
6	4	15 3	5	6 2	1	14 8	9 7
1	3	8	17 7	4	50 5	6	2
35 7	6 6	1	4	168 8	2	5	12 3
5	84 7	2	6	3	9 8	1	4
32 4	7 5	6	7 8	7	19 3	2	1
8	2	28 7	1	24 6	4	21 3	5
2 2	1	4	90 3	5	6	7	8

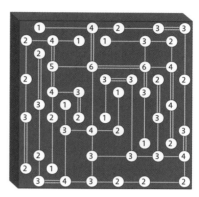

52

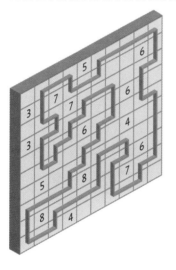

48

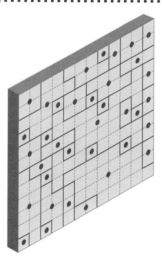

54

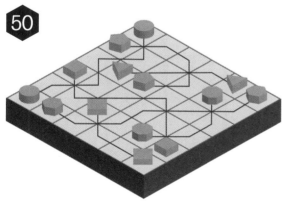

50

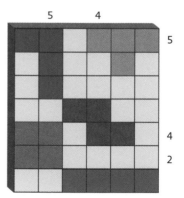

56

121

SOLUTIONS

58

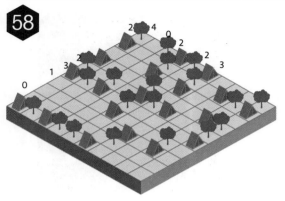

64

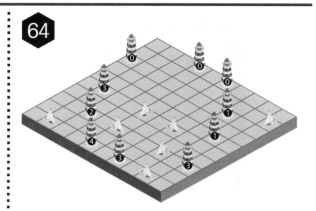

60

35	34	33	31	30	26	25	24	18	17
38	36	99	32	27	29	23	20	19	16
39	37	98	100	28	22	21	4	15	14
40	96	97	90	89	1	3	5	12	13
41	94	95	91	88	2	6	7	10	11
42	45	93	92	87	86	8	9	68	67
44	43	46	56	84	85	60	71	69	66
49	47	55	83	57	59	72	61	70	65
48	50	82	54	58	73	74	77	62	64
51	52	53	81	80	79	78	75	76	63

65

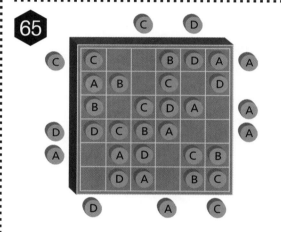

62

61 = 29 + 11 + 8 + 13

67 = 16 + 9 + 8 + 34

117 = 39 + 22 + 19 + 37

67

69

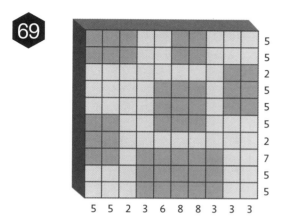

Column clues (right side, top to bottom): 5, 5, 2, 5, 5, 5, 2, 7, 5, 5

Column clues (bottom, left to right): 5, 5, 2, 3, 6, 8, 8, 3, 3, 3

75

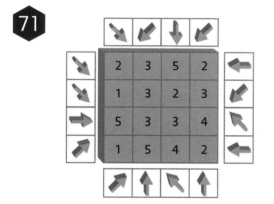

9	3	5	7	1	6	4	8	2
4	9	3	2	8	7	6	1	5
3	2	1	8	4	9	5	6	7
5	8	6	1	7	4	2	3	9
1	7	2	5	6	8	3	9	4
6	4	8	3	9	2	7	5	1
7	6	9	4	5	3	1	2	8
8	1	7	6	2	5	9	4	3
2	5	4	9	3	1	8	7	6

71

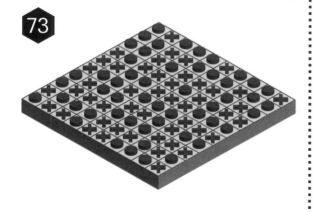

2	3	5	2
1	3	2	3
5	3	3	4
1	5	4	2

77

Top clues: 4, 2, 3, 5

7	2	5	6	3	4	1	4
1	3	4	7	6	5	2	
2	6	7	5	4	1	3	4
5	7	3	4	1	2	6	2
3	5	6	1	2	7	4	
6	4	1	2	5	3	7	
4	1	2	3	7	6	5	

Left clues (rows): 4, 4, 4, 2

Bottom clues: 4, 3, 4, 1

73

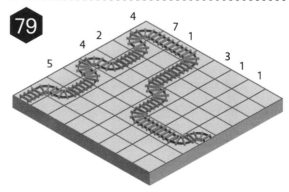

79

Clues around grid: 5, 4, 2, 4, 7, 1, 3, 1, 1

SOLUTIONS

66

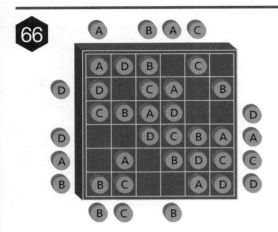

72

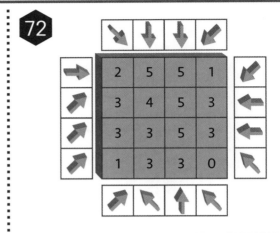

68

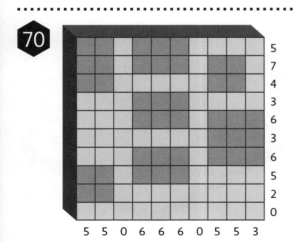

74

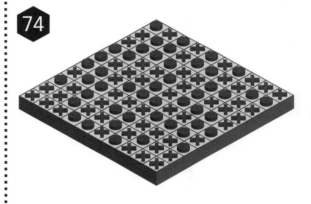

70

										5
										7
										4
										3
										6
										3
										6
										5
										2
										0
5	5	0	6	6	6	0	5	5	3	

76

6	2	1	8	9	7	4	3	5
2	4	8	7	1	6	3	5	9
1	3	9	5	7	4	2	8	6
8	1	2	6	3	5	9	7	4
5	9	7	3	4	2	1	6	8
4	8	6	1	2	3	5	9	7
3	6	5	2	8	9	7	4	1
9	7	3	4	5	8	6	1	2
7	5	4	9	6	1	8	2	3

78

Top clues: 2 5 2 5 2

	6	7	1	3	5	2	4	3
2	5	4	2	7	6	3	1	
	2	1	5	6	7	4	3	
6	1	2	3	5	4	6	7	
	7	5	6	4	3	1	2	5
2	4	3	7	2	1	5	6	
	3	6	4	1	2	7	5	

Bottom clues: 3   6 4

80

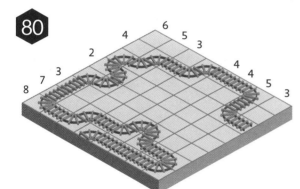

81

7	2	1	8	4	5	6	
	1	3	6	5	2	8	4
3	4	8	2		6	1	7
4	3		5	2	1	7	8
5	6	4		3	7	2	1
8		6	4	1	3	5	2
6	8	5	1	7		4	3
1	7	2	3	8	4		5

83

4   2
4   3   4
4 7 4 2 3 6 6 5
5   8   2   1
4 3

85

²4	6	⁷⁵5	²2	²¹7	3	⁹1	8
¹⁹⁶⁰7	5	3	4	⁸1	8	¹²2	6
5	7	8	¹²6	2	¹²⁰1	4	²¹3
⁵8	3	²⁰2	1	4	5	6	7
³6	¹²⁸4	1	3	8	2	¹²7	5
2	8	4	¹²7	5	³⁶6	3	1
⁴3	1	²7	5	¹⁸6	¹⁹4	8	2
¹1	2	⁴⁸6	8	3	7	²⁰5	4

87

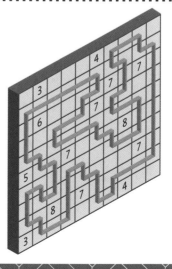

4   7
3
7   7
6   7   8
7
5   7
7   4
8
3

SOLUTIONS

89

95

91

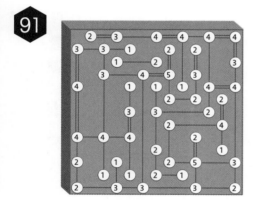

97

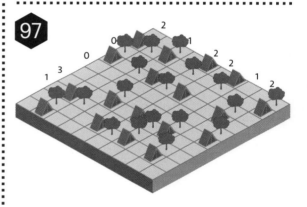

93

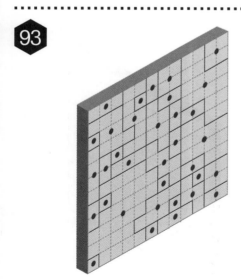

99

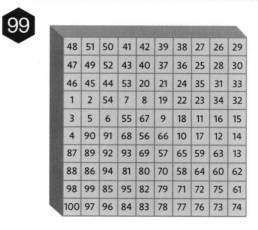

101

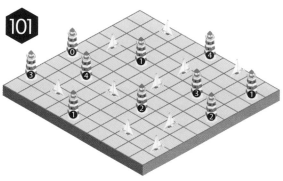

82

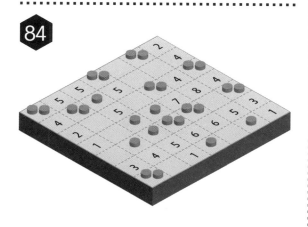

84

86

21 7	2 2	96 8	6	1	1 4	90 5	3
3	1	42 7	2	19 8	5	6	6 4
14 1	11 5	6	14 4	3	8	21 7	2
8	6	4	1	280 7	2	3	240 5
5	28 7	2	3	4	1	10 8	6
24 6	4	16 1	7	5	2 3	2	8
4	48 3	15 1	5	420 2	6	4 1	7 7
2	8	3	5	6	7	4	1

88

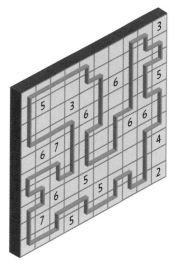

90

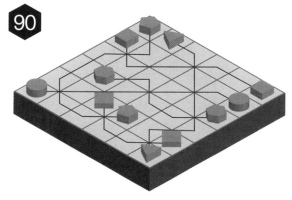

127

SOLUTIONS

92

98

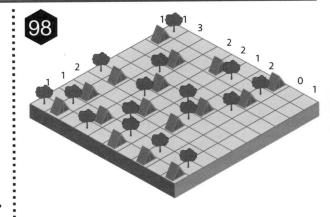

94

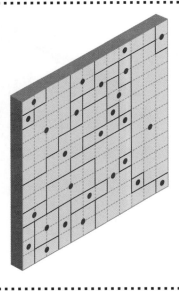

100

59	57	56	55	54	39	40	25	24	23
58	60	61	62	53	41	38	26	22	21
48	50	46	52	63	37	42	29	27	20
49	47	51	45	64	43	36	28	30	19
6	4	3	65	44	35	16	17	18	31
5	7	2	13	66	15	34	33	32	84
8	10	12	1	14	67	87	80	85	83
9	11	91	90	68	88	79	86	81	82
97	99	95	92	89	69	72	78	77	75
98	96	100	94	93	71	70	73	74	76

96

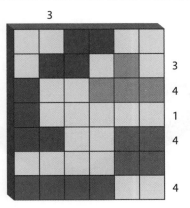

102

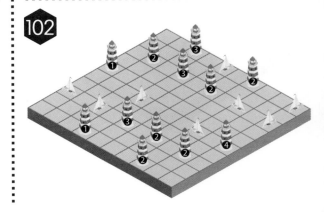

103

77 = 18 + 12 + 33 + 14

79 = 17 + 30 + 9 + 23

129 = 38 + 21 + 33 + 37

 105

107

 109

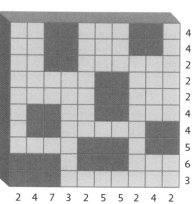

111

113

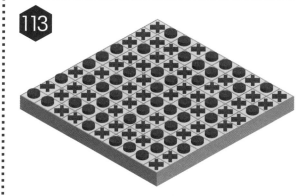

SOLUTIONS

115

9	8	7	2	3	1	4	6	5
5	4	6	8	2	7	1	9	3
4	5	8	9	6	3	7	1	2
1	7	3	4	9	5	6	2	8
2	6	9	7	4	8	3	5	1
3	1	5	6	8	9	2	4	7
7	3	2	1	5	6	9	8	4
6	2	1	5	7	4	8	3	9
8	9	4	3	1	2	5	7	6

121

3		4	1	5	2	8	6
7	4	3		2	1	6	8
5	1	6	8	3	7	2	
2	5	8	3		6	7	1
	2	7	6	1	4	5	3
4	6		5	7	8	3	2
1	8	5	2	4	3		7
6	7	2	4	8		1	5

118

			4	4	5		
	7	5	6	4	3	2	1	6
3	3	6	5	2	4	1	7
3	1	4	7	5	6	3	2	4
	6	7	2	3	1	5	4
4	2	1	3	6	7	4	5	2
	4	2	1	7	5	6	3
2	5	3	4	1	2	7	6

Bottom: 3 2 2 2 3

119

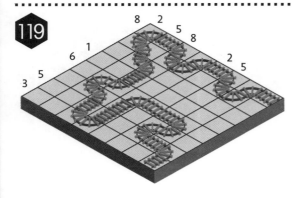

123

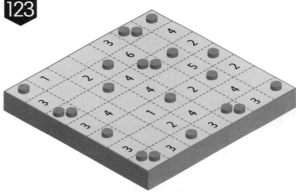

125

12 2	6	13 4	3	13 5	18 1	56 8	11 7
15 5	1	3	6	8	2	7	4
1	56 7	8	2	4	6	3	12 5
224 8	7 3	8 2	5 5	1	35 7	6 4	6
7	4	6	11 8	3	5	2	1
4	350 5	7	1	2	14 8	6	480 3
18 3	16 2	1	11 7	126 6	4	5	8
6	8	5	4	7	3	2 1	2

127

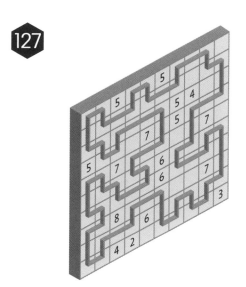

133

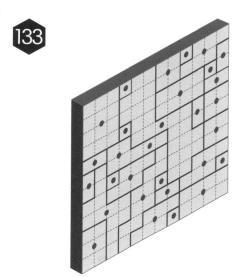

129

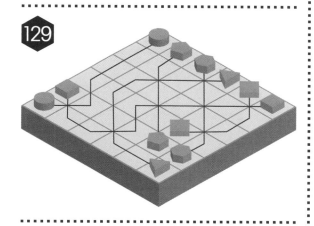

135

131

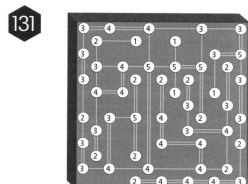

137

SOLUTIONS

10	8	6	5	4	3	2	1	68	67
11	9	7	81	73	72	70	69	66	65
12	13	82	83	80	74	71	61	63	64
15	14	19	84	79	77	75	62	60	57
16	18	20	85	86	78	76	59	58	56
17	21	94	91	92	87	48	50	55	54
22	95	97	93	90	88	49	47	51	53
23	24	96	98	99	89	40	42	46	52
25	27	30	31	100	34	41	39	43	45
26	28	29	32	33	35	36	37	38	44

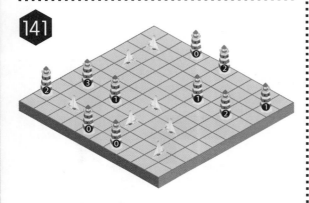

104

77 = 8 + 26 + 24 + 19

100 = 32 + 27 + 24 + 17

131 = 32 + 29 + 34 + 36

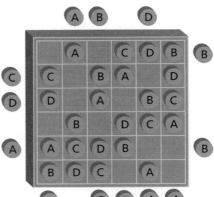

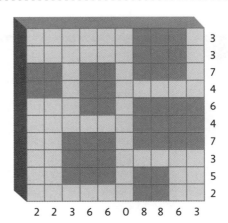

112

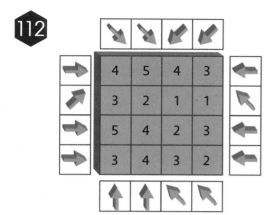

117

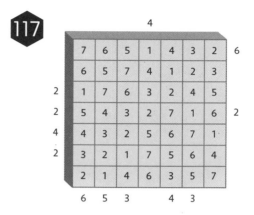

114

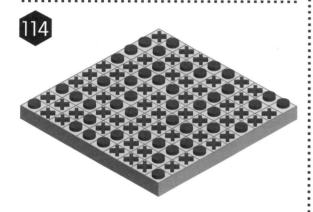

120

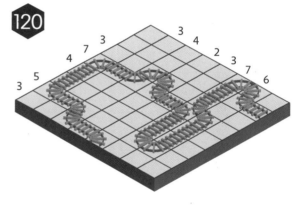

116

9	8	5	3	2	1	4	7	6
2	3	9	7	5	8	6	4	1
6	7	1	4	9	2	5	8	3
8	4	7	6	1	5	3	9	2
7	1	4	9	3	6	8	2	5
1	5	3	2	8	4	9	6	7
3	6	8	1	7	9	2	5	4
4	9	2	5	6	7	1	3	8
5	2	6	8	4	3	7	1	9

122

4		8	7	6	5	1	2
6	4	2	8	3	1	5	
7	1	3	5	8	4		6
8	3	5		2	7	6	1
	7	1	4	5	3	2	8
3	5	6	1		2	7	4
5	2	7	6	1		4	3
1	8		2	4	6	3	5

SOLUTIONS

124

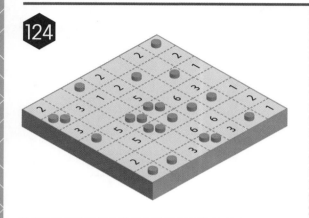

130

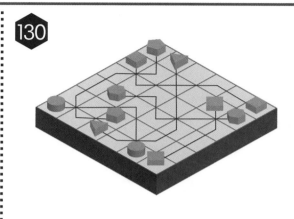

126

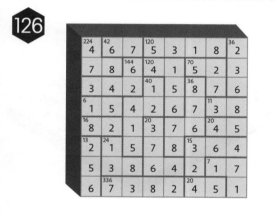

132

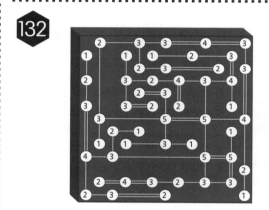

128

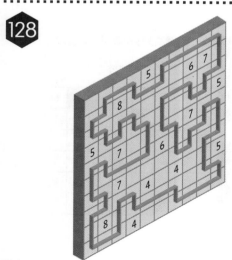

134

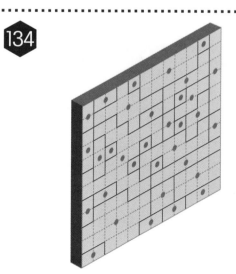

136

142

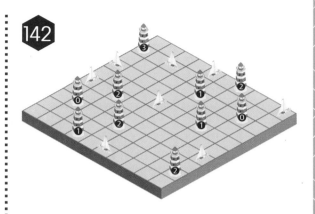

138

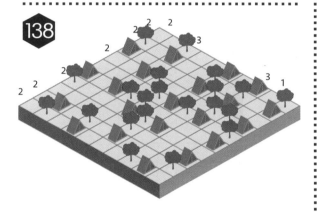

143

74 = 19 + 22 + 18 + 15
91 = 14 + 20 + 30 + 27
129 = 40 + 22 + 38 + 29

145

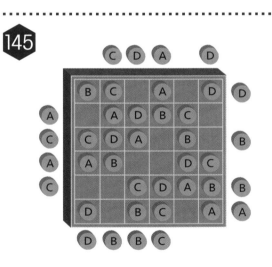

140

43	39	38	37	36	29	27	26	100	99
44	42	40	35	30	28	24	25	96	98
45	46	41	34	31	23	21	95	92	97
50	51	47	32	33	22	20	93	94	91
52	49	48	73	74	17	18	19	89	90
53	54	72	75	76	81	16	88	87	86
70	71	55	79	80	77	82	15	84	85
68	69	56	57	78	13	14	83	5	6
67	63	64	60	58	12	2	3	4	7
66	65	62	61	59	1	11	10	9	8

135

SOLUTIONS

147

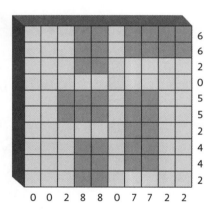

153

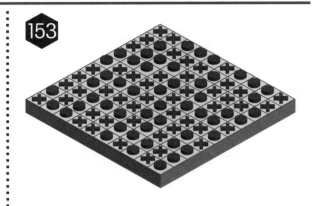

149

	6
	6
	2
	0
	5
	5
	2
	4
	4
	2

0 0 2 8 8 0 7 7 2 2

155

7	4	6	9	5	1	8	3	2
6	9	2	5	8	3	1	4	7
9	8	3	2	6	7	4	1	5
3	1	9	7	4	6	2	5	8
5	2	1	8	3	4	7	9	6
4	3	5	1	7	8	6	2	9
8	5	7	3	1	2	9	6	4
1	7	4	6	2	9	5	8	3
2	6	8	4	9	5	3	7	1

151

4	3	4	5
3	1	1	2
5	2	1	3
4	1	1	2

157

		3		2		2	
	7	1	3	2	4	5	6
	6	7	2	3	1	4	5	2
4	3	4	1	5	7	6	2
4	1	2	6	4	5	7	3
3	4	3	5	7	6	2	1	4
3	2	5	7	6	3	1	4	3
	5	6	4	1	2	3	7

3 2 2 4 2

159

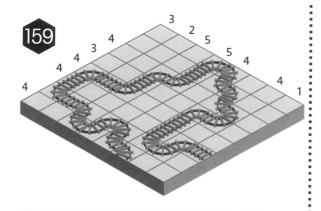

144

59 = 23 + 12 + 9 + 15
104 = 36 + 12 + 40 + 16
105 = 19 + 32 + 40 + 14

146

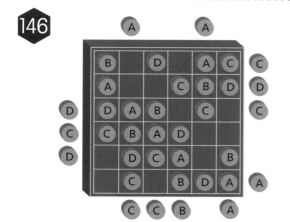

148

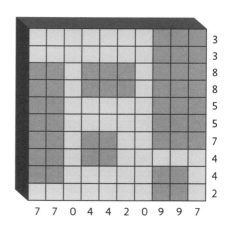

150

152

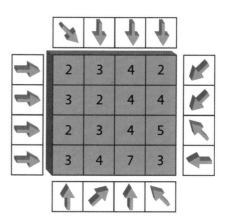

SOLUTIONS

154
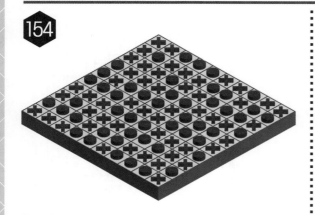

160
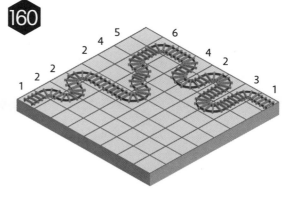

156

8	4	7	3	5	6	2	9	1
2	1	8	5	3	4	9	7	6
7	6	9	4	2	1	8	3	5
9	8	3	1	6	2	5	4	7
5	9	6	7	4	8	3	1	2
1	3	2	9	8	5	7	6	4
6	5	1	8	9	7	4	2	3
4	2	5	6	7	3	1	8	9
3	7	4	2	1	9	6	5	8

161
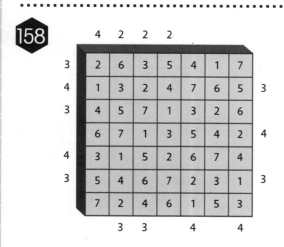

158

	4	2	2	2			
3	2	6	3	5	4	1	7	
4	1	3	2	4	7	6	5	3
3	4	5	7	1	3	2	6	
	6	7	1	3	5	4	2	4
4	3	1	5	2	6	7	4	
3	5	4	6	7	2	3	1	3
	7	2	4	6	1	5	3	
		3	3	4		4		

164
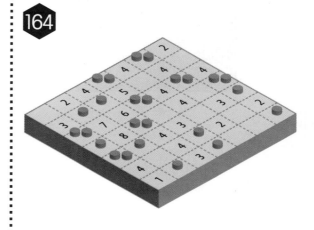

166

72 3	6	2	56 7	20 4	210 1	5	9 8
2	13 3	6	8	5	32 4	7	1
35 7	5	4	6 2	3	8	1	6
20 4	8	5	28 1	7	180 3	4 6	2
56 8	7	3	4	1	6	2	5
13 5	1	56 7	14 6	8	40 2	12 3	4
6 1	2	8	8 3	12 6	5	4	25 7
6	4	1	5	2	7	8	3

171

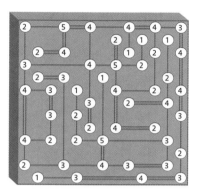

167

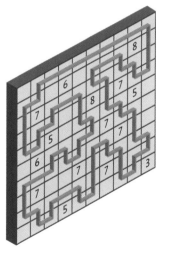

174

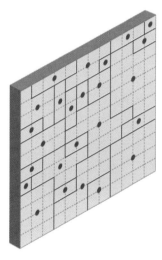

169

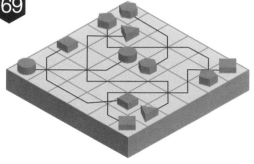

175

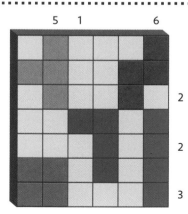

SOLUTIONS

177

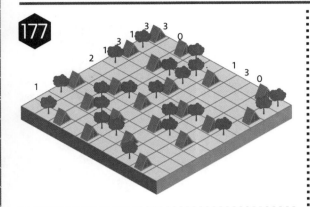

183

$70 = 23 + 18 + 16 + 13$

$73 = 19 + 18 + 9 + 27$

$118 = 32 + 40 + 16 + 30$

162

3	7	6	2	1		5	4
4	6	3	5	2	1		7
6		5	8	7	2	3	1
7	5		3	8	4	2	6
1	3	8		5	6	4	2
2	4	7	6	3	8	1	
8	1	2	4		5	6	3
	2	4	1	6	3	7	5

179

28	17	16	19	12	11	44	43	46	47
29	27	18	15	20	13	10	45	42	48
30	26	32	21	14	9	8	51	49	41
86	31	25	33	22	7	52	50	40	39
87	85	24	23	34	6	36	53	38	55
90	88	84	82	99	35	5	37	54	56
89	91	83	98	81	100	4	61	57	59
92	93	97	80	71	1	62	3	60	58
94	96	75	72	79	70	2	63	64	65
95	74	73	76	77	78	69	68	67	66

163

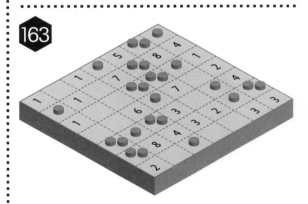

165

29 8	2	4	6	1 7	1	45 5	3
1050 5	6	1	8	2	32 4	3	112 7
7	5	6 6	1	3	2	4	8
1	336 4	3	7	1920 8	30 5	6	2
24 2	21 3	7	4	5	6	8	48 1
3	21 8	5	12 2	13 6	7	1	4
4	42 7	8	5	28 1	3	2	6
6	1	2	3	4	8	7	5

181

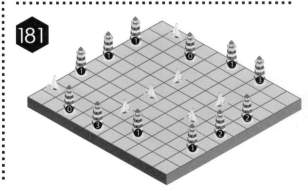

168

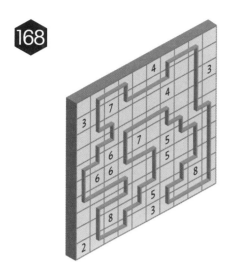

173

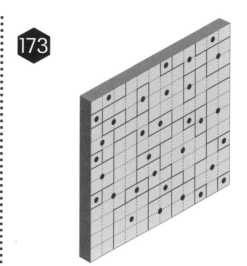

170

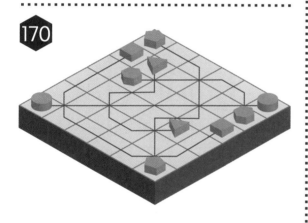

176

172

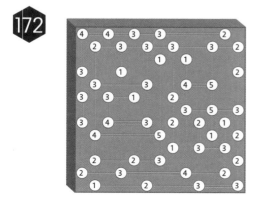

178

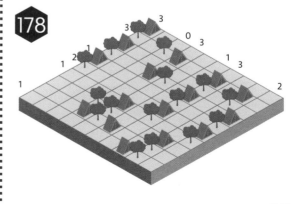

141

SOLUTIONS

180

99	98	87	86	85	83	17	18	6	5
100	97	88	89	84	82	16	7	19	4
95	96	90	91	81	15	8	20	1	3
94	93	92	80	14	12	9	21	32	2
74	73	79	39	13	10	11	22	33	31
72	75	78	40	38	36	35	34	23	30
71	76	77	43	41	37	25	24	27	29
66	70	69	42	44	45	46	26	48	28
65	67	68	60	59	57	55	47	51	49
64	63	62	61	58	56	54	53	52	50

182

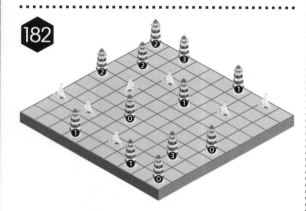

184

81 = 24 + 10 + 20 + 27
99 = 14 + 12 + 34 + 39
111 = 35 + 18 + 20 + 38

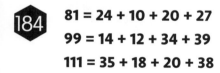

200

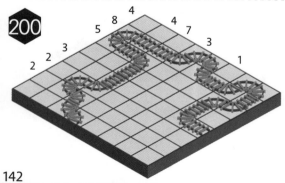

185

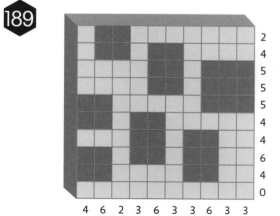

187

7	9	8	3	6	5	2	1	4
3	5	4	7	2	1	6	8	9
2	6	1	9	4	8	3	5	7
6	2	3	8	5	9	7	4	1
5	8	7	2	1	4	9	3	6
1	4	9	6	3	7	5	2	8
9	3	2	4	8	6	1	7	5
8	7	5	1	9	3	4	6	2
4	1	6	5	7	2	8	9	3

189

191

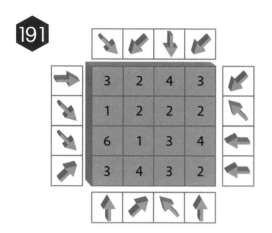

193

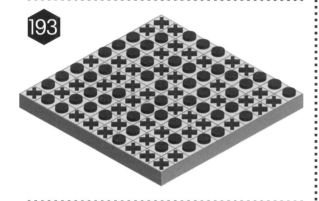

195

6	2	5	9	1	3	4	7	8
5	4	2	7	8	1	9	3	6
1	9	4	5	7	2	6	8	3
4	1	8	6	3	9	2	5	7
9	5	1	3	6	7	8	2	4
2	6	7	4	5	8	3	9	1
8	7	3	2	4	6	5	1	9
3	8	9	1	2	4	7	6	5
7	3	6	8	9	5	1	4	2

198

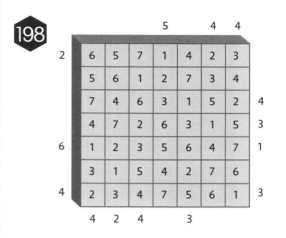

199

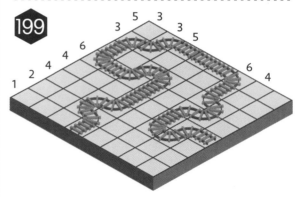

186

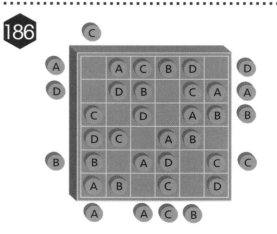

SOLUTIONS

188

1	8	6	7	4	3	5	9	2
2	7	3	8	5	9	1	6	4
5	4	9	2	6	1	7	8	3
7	6	4	5	1	8	2	3	9
8	3	5	4	9	2	6	1	7
9	1	2	6	3	7	4	5	8
3	5	8	1	7	4	9	2	6
4	2	1	9	8	6	3	7	5
6	9	7	3	2	5	8	4	1

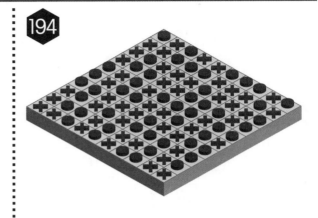**194**

190

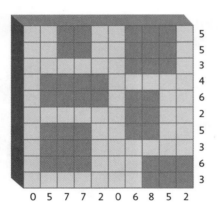

5 5 5 3 4 6 2 5 3 6 3

0 5 7 7 2 0 6 8 5 2

196

9	1	3	7	5	8	4	2	6
1	4	9	2	6	5	3	7	8
8	6	2	9	4	7	5	1	3
7	5	6	4	2	1	8	3	9
3	9	8	1	7	4	2	6	5
6	2	7	5	1	3	9	8	4
5	7	4	3	8	6	1	9	2
4	3	1	8	9	2	6	5	7
2	8	5	6	3	9	7	4	1

192

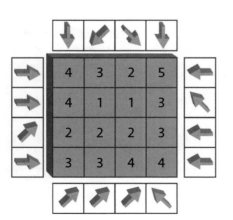

	4	3	2	5	
	4	1	1	3	
	2	2	2	3	
	3	3	4	4	

197

	5	4		4	3		
1	7	2	3	6	1	4	5	3
4	4	3	5	1	2	6	7	
	6	5	2	7	4	3	1	4
3	3	6	4	5	7	1	2	2
3	5	4	1	2	6	7	3	
	1	7	6	3	5	2	4	4
2	2	1	7	4	3	5	6	
	4		1	3	4			